TEACHERS' LEARNING

Science & Technology Education Library

VOLUME 7

SCOPE

The book series *Science & Technology Education Library* provides a publication forum for scholarship in science and technology education. It aims to publish innovative books which are at the forefront of the field. Monographs as well as collections of papers will be published.

Teachers' Learning

Stories of Science Education

by

JOHN WALLACE

*Curtin University of Technology,
Perth, Australia*

and

WILLIAM LOUDEN

*Edith Cowan University,
Perth, Australia*

KLUWER ACADEMIC PUBLISHERS

DORDRECHT / BOSTON / LONDON

A C.I.P. Catalogue record for this book is available from the Library of Congress.

ISBN 0-7923-6259-4 (HB)

Published by Kluwer Academic Publishers,
P.O. Box 17, 3300 AA Dordrecht, The Netherlands.

Sold and distributed in North, Central and South America
by Kluwer Academic Publishers,
101 Philip Drive, Norwell, MA 02061, U.S.A.

In all other countries, sold and distributed
by Kluwer Academic Publishers,
P.O. Box 322, 3300 AH Dordrecht, The Netherlands.

Printed on acid-free paper

Printed in the Netherlands.

TABLE OF CONTENTS

FOREWORD

Research on science teacher education has rapidly increased in quantity and improved in quality over the last two decades. Changes have been characterised by improvements in research methods, increases in the range of alternative methodologies used and expansion of the diversity of the theories applied to research methods and descriptions of critical issues associated with teaching, learning, and teacher learning. Throughout my own research career I have been involved personally in many of the changes that have occurred and have experienced first hand the difficulty of gaining acceptance for alternatives to the tried and tested approaches of the past. It has been an exciting period in which to undertake research in science education and teacher education as theories and practices from a diverse set of social sciences have found their way into the praxis of educational researchers. In *Teachers' Learning: Stories of Science Education* John Wallace and Bill Louden exemplify these changes. The authors employ narrative methods in a series of case studies that explore the teaching and learning of science and the manner in which teachers learn and change the enacted curriculum. In so doing they build on their experiences as graduate students at the University of Toronto's Ontario Institute for Studies in Education (OISE) and on pioneering applications of narrative in research on teaching (e.g., Connelly & Clandinin, 1985). Some of the case studies in the book are grounded in the authors' experiences at OISE and others derive from their experiences as teachers, teacher educators and educational researchers in Australia.

Through the use of the stories of teachers Wallace and Louden reject some of the traditional approaches to research in which narratives are prepared as texts for codification. A variety of vignettes portray teacher knowledge as dynamic and grounded in experience as teachers overcome problems while enacting curricula in their classrooms. The significance of opportunities for reflection is stressed throughout the book and the associated value of collaboration with others is emphasised as a critical factor in teachers learning to teach and changing their patterns of practice.

Telling stories based on personal experiences allows teachers and researchers to re-present what they have learned in holistic ways that reflect their biographies and the perspectives used to give meaning to their praxis. Acknowledged subjectivity and a reticence of the researchers to tell readers what they should learn and generalise are strengths of the approaches adopted by Wallace and Louden. Without exception the studies show evidence of a close relationship between the researchers and the researched. Inevitably ethical issues arise and the chapters serve to show readers not only how to report studies through the use of narrative but also how to treat teachers with respect and to re-present their perspectives on teaching and learning in ways from which readers can learn. The approach avoids the all-too-familiar violence inflicted by researchers on practitioners when publications are used to criticise the extant practices of teachers.

Teacher education programs make numerous assumptions about the nature of knowledge and how it is acquired. What are the best ways to learn to teach? How is knowledge of teaching applied to improve teaching? Too often questions such as these are not asked by teacher educators. Instead answers to these and other unasked questions are taken for granted and are incorporated in the enactment of degree and non-degree courses for practising and prospective teachers. For example, it is common to find teacher education programs maintaining traditional divisions between science teaching in elementary and secondary classrooms and relying on coursework that is typically carried out in university classrooms and assumes that learning to teach occurs by talking, listening, writing and reading. Too frequently teacher education courses assume that knowledge derived from the study of books and journal articles can be applied while teaching. It is not surprising that teachers often complain about the lack of relevance of traditional courses in teacher education and that teacher educators struggle to bridge gaps between problems of practice and courses for teachers. Through detailed case studies Wallace and Louden show that what is learned about teaching and the process of change is context dependent and what happens in classrooms extends beyond what teachers believe and can talk and write about. Context is significant in shaping what teachers learn from their practices, how they enact curricula, and the extent to which desired changes occur. The case studies show the significance of the biographies of teachers and contextual factors such as whether or not the teachers are in or out of field, the expectations and roles of students, the actions of peers and school administrators, and the presence of policies that exhort reform.

As a set the case studies provide food for thought for teachers and teacher educators embarking on projects of change, whether their concerns and goals are local and focused on improvement within a single class or are more systemic and concerned with reform throughout one or more schools. The studies transcend elementary and high school grade levels and, although the primary focus is on the teaching and learning of science, *Teachers' Learning: Stories of Science Education* has potential appeal to prospective and practising teachers in diverse subject areas, policy makers, teacher educators and researchers. Wallace and Louden have added a significant title to the Science and Technology Education Library, one that is likely to make an impact on the fields of science education and teacher education.

Kenneth Tobin
Professor
University of Pennsylvania
Philadelphia, PA 19104-6216, USA

SECTION I

INTRODUCTION

1. STORIES AND SCIENCE

This is a book about teachers and about science teachers in particular. In writing this book we have been preoccupied with a series of related questions about how teachers learn to teach science. What is the nature of the knowledge teachers have and use? How do teachers grow and change? How does subject content and school context influence teachers' learning? What role does reflection play in teachers' learning? What forms of collaboration best support teachers' learning? For more than a decade we have been working through our preoccupations with these issues in the usual ways of researchers: visiting classrooms, watching and participating in lessons, interviewing teachers and students, and working with teachers in workshops and higher degree studies. The body of work that has accumulated has at least one thing in common: all of the research reported in this book uses stories written by or about teachers. These stories — which might more formally be called narratives, case studies or vignettes — serve for the most part as a form of data for arguments we want to make about the topic that has preoccupied us: teachers' learning.

Because 'stories' and 'science' might seem to be an odd couple, with stories imagined as a soft form of knowledge and science imagined as the harder, more rigorous and more reliable of the pair, we begin with a brief exploration of our understanding of the relationship between stories and science.

One kind of account of the relationship between stories and science is that they are different ways of knowing, best suited to different kinds of problems. Schön, for example, distinguished between the high ground of technical rationality and the low ground, the 'indeterminate swampy zones of practice' (Schön, 1987, p. 3). Similarly, Bruner has characterised science as one of a pair of alternative ways of knowing: *paradigmatic* ways of knowing and *narrative* ways of knowing. What he calls paradigmatic ways of knowing are analytic, general, abstract, impersonal and decontextualised. In contrast, what he calls narrative ways of knowing

3

are specific, local, concrete, personal and contextualised (Bruner, 1986).
Like C. P. Snow's 'two cultures' (1964) or the commonplace distinction
between arts and sciences, this kind of dichotomy associates the scientific
way of knowing with the Enlightenment project of reason overcoming
superstition and blind faith.

An alternative strategy is to contest the distinction between narrative
and science; to argue that science is just another narrative. In *The
Postmodern Condition*, the philosopher Lyotard begins his discussion
of the legitimation of scientific knowledge by arguing that science has
always seen itself in conflict with, and superior to, narrative (Lyotard,
1984). The strategy of science, he says, has been to locate itself as the
one true source of authorised knowledge. In the service of the
Enlightenment vision of reason, science has developed canons of
evidence, patterns of discourse and forms of reasoning which are argued
to be superior to the knowledge held in commonsense and narrative
accounts of the world. Refusing the special privilege science has reserved
for itself, Lyotard characterises science as a metanarrative, a grand and
inclusive story which offers to provide complete explanations for the
phenomena it addresses. Like the other great metanarratives of the
nineteenth and twentieth centuries (Marxism and psychoanalysis), science
overstates the capacity of its methods to yield up the truth, and overstates
its superiority over the personal, local and contextual knowledge held in
smaller stories (*petit récits*). Summarising his views on the state of
knowledge in the late twentieth century, Lyotard argues that the defining
characteristic of the postmodern condition is 'incredulity towards
metanarratives' such as science (Lyotard, 1984, p. xxiv).

The notion that 'science, too, is a story' (Feyerabend, 1995, p. 103) is
at the heart of the contemporary legitimation crisis in science. In this
context, it is worth remembering that the scientific method has a short,
recent and local history. The Enlightenment vision of science is just a
few hundred years old, and it has arisen and flowered in just one of the
world's rich range of cultures. Scientific constructions of truth may have
been naturalised in the popular imagination in the West, but they also
have been subjected to a series of destabilising philosophical critiques.
Kuhn's (1962) critique has been understood as arguing that what counts
as the truth in science is no more than intersubjective agreement among
a group of scientists, a kind of fashion in thinking which may be swept

away by the fashion of next season. Feyerabend's trenchant *Against Method* (1975) extended this critique, arguing that scientists do not solve problems because they follow the scientific method but because 'they have studied a problem for a long time, because they know the situation fairly well, because they are not too dumb . . . and because the excesses of one scientific school are almost always balanced by the excesses of some other school' (Feyerabend, 1975, p. 302).

Scepticism about the capacity of the scientific method to underwrite truth claims has been fuelled by philosophers such as Foucault (1977) who argued that knowledge is inseparable from power. What counts as knowledge, according to Foucault, is the result of patterns of power relations in a community: there is no power relation which does not produce a field of knowledge, and no field of knowledge which is not constituted by its power relations (1977, p. 27). Science, like any other system of knowledge, is embodied in systems of discourse which 'systematically form the objects about which they speak' (Foucault, 1972, p. 49). These discourses, such as the discourse of science, construct and naturalise 'truths' which allow for the regulation of persons in the discourse community. Reversing Francis Bacon's dictum that 'knowledge itself is power', Foucault concludes that power creates knowledge.

Taken together, the insights of philosophers of science such as Kuhn, postmodernist philosophers such as Lyotard and Feyerabend and poststructuralist cultural critics such as Foucault, have led to the end of certainty that science is a separate way of knowing, superior to narrative. Rather than providing a method which securely underwrites truth claims, science has been characterised as a form of knowledge constructed and agreed among like-minded groups of people, reflecting the power structures within which the knowledge is created, likely to change when the power structures change, and sceptical of truth claims arising outside its own Enlightenment metanarrative of a triumphant reason.

For educational researchers, among the implications of this critique of the metanarrative of science has been the end of certainty about research methods. One response to this uncertainty has been a long series of qualitative/quantitative arguments carried out in the journals of the American Educational Research Association (see, for example, Loving, 1997). The nearing end of the paradigm wars in that journal was signalled in an article protesting against 'the hegemony of the narrative' (Cizek,

1995). An alternative response has been the creation of a whole new cottage industry of handbooks and guides to qualitative research, aimed at showing how the rigour of research may be guaranteed in a postmodern era. Beginning with the landmark work of Glaser and Strauss (1967) and Guba (1978) the long series of such works includes Guba and Lincoln (1981), Goetz and LeCompte (1984), Lincoln and Guba (1985), LeCompte, Millroy and Preissle (1992), Miles and Huberman (1993), and Denzin and Lincoln (1994).

Such handbooks have proven very useful to postgraduate students because they offer secure prescriptions about method. In a world after reliability and validity, the handbooks offer criteria of trustworthiness such as 'credibility', 'transferability', 'dependability' and 'confirmability' (Lincoln & Guba, 1985, p. 189). Because we recognise that dissertation examiners and journal reviewers need to feel confident that the work they are reading has been constructed in relation to some kind of standards, we recommend the handbooks to our students. We do not, however, believe that the method handbooks overcome the problem of method, which is that no method can guarantee the truth in a postmodern world. Whatever emerges from a program of disciplined inquiry must be constructed within a web of intersubjective agreement, reflecting the preconceptions of the authors and the power structures within which the knowledge is constructed.

Methodologically, we see this book as 'an analysis of narratives', to use Polkinghorne's (1995) distinction. That is, we use stories as a data source for more general arguments about teachers' learning. In doing so, we are aware of the instability of arguments we have constructed around these stories. In the discussion that follows, we identify four sources of instability in analysis of narratives and position our research in terms of these sources of instability: the borderlands of fact and fiction, the problem of authenticity, the never ending story of interpretation, and the problem of generalisation.[1]

THE BORDERLANDS OF FACT AND FICTION

All of the stories in this book are based on some kind of disciplined encounter with students, teachers and schools. We collected the data for most of the chapters in school visits. We wrote ethnographic field notes, interviewed teachers and looked at the documentary records teachers keep. For some chapters, the stories were created by teachers or postgraduate students on the basis of their own experience in classrooms. Notwithstanding this basis in fact, the stories are all 'made up' in the sense that they involve selection and juxtaposition of data and that they involve layers of interpretations about the events described. As Clifford has said, such ethnographic narratives are 'constructed domains of truth — serious fictions' (Clifford, 1988, p. 100).

To think of ethnographic stories as 'serious fictions' is to locate them in a borderland which includes texts which are offered as works of imagination, fictionalised treatments of real events and research works presented in narrative form. From the perspective of a library cataloguer, these are not borderlands but an iron curtain; a work is fiction or non-fiction and the classification is spelt out on the spine. But the breadth of the borderlands is illustrated by two recent books: Garner's *The First Stone* (1995) and Schama's, *Dead Certainties* (1991). *The First Stone* is a novelist's inquiry into a notorious case of sexual harassment at an elite residential college attached to an Australian university. The events she investigated were entirely real: an end of term dance, a late-night encounter between a young female student and the college Master, the trial and acquittal of the Master, and Garner's own involvement in a bitter debate among feminists about the events. In this regard, the work is clearly a work of non-fiction; what is less typical is the partisan role Garner played in the events and the fictive techniques she used to tell the story. Instead of the hard edged facticity which might be expected of a piece of investigative journalism, *The First Stone* is narrated in the same intensely introspective first person that Garner has used in her novels. Her willingness to connect the issues of harassment with her own experience as a second-wave libertarian feminist in the 1970s produced two broad kinds of reaction. Some, whose commitments were offended by Garner's lack of sympathy for the complainants, critiqued Garner's method (Pybus, 1995); others, including those who had previously admired her fiction, made the book an instant best seller.

Dead Certainties traverses the same borderlands, moving in the opposite direction. Schama is an award winning writer of academic histories. His 1989 book *Citizens* is preceded with the warning that

'Though it is in no sense fiction (for there is no deliberate invention) it may well strike the reader as story rather than history' (Schama, 1989, p. 6). In *Dead Certainties* he experimented more boldly with alternative forms of historical narrative, creating deliberate and acknowledged fictions in his study of the uncertainty around accounts of the lives and deaths of two historical characters. Like Garner's work, there is fact in the events which may be checked against sources, and there is fiction in the imaginative way Schama's story is told and in his willing acknowledgement that 'claims to historical knowledge must always be circumscribed by the character and prejudices of its narrator' (Schama, 1991, p. 322).

In this book, the stories are all true in the sense that 'there is no deliberate invention'. They are, however, more personal than much of the research published in the field of science education. Chapter 2, for example, follows one of us (Bill) through a cycle of collaborative teaching and learning with a colleague and includes reports of failure as well as success. Similarly, in Chapter 6 the discussion of ethics in narrative research uses stories about ethical dilemmas we have both encountered and solved to greater or lesser degrees of satisfaction. Throughout the book, one or other of us is always present as an observer and storyteller in a first person narrative, always involved in the events in ways that embody the preconceptions we bring to the study of teachers' learning. We take the view that people's understanding of events is always constructed through the preconceptions they bring to the act of understanding. There is no prior state of understanding free of our preconceptions, and no method which can free readers and writers from the preconceptions they bring to each new experience (Gadamer, 1975). Consequently we argue that it is important to pay particular attention to the influence of preconceptions on other participants' understanding of events, and to the influence of our own preconceptions on our understanding of these events. This view of the constitutive role of preconceptions in formation of knowledge may not suit researchers trained to regard their own preconceptions as a threat to validity and reliability. Be that as it may, it also means that we distrust the idea that following a predetermined method is a reliable path to the truth. No method, no matter how clearly it is represented in a Handbook, can free researchers from their preconceptions or deliver them an incontestable truth. No method can free researchers from having to account for the constructions of reality they make as they follow any research method and produce a written account of the interpretations they make.

THE PROBLEM OF AUTHENTICITY

The idea that truth claims are 'circumscribed by the character and prejudices of the narrator' brings to narrative research the problem of authenticity. Since the narrator's voice is implicated in truth claims, readers are entitled to know something about the authenticity of the narrator's voice. In Australia, the issue of voice has been focused in a series of recent fictions. The most notorious of these concerns the novel *The Hand that Signed the Paper* by an author named Helen Demidenko (1995). This first novel, which won three of Australia's most prestigious literary awards, provides an account of the Holocaust from the point of view of Ukrainian collaborators. For more than a year the author represented herself as the daughter of Ukrainian immigrants, and the stories in the novel as authentic and untold family history. She addressed writers' festivals in folk costumes, persuaded immigrant Ukrainian academics of her authenticity (Longley, 1997), and 'distracted the literati with pretty folk dances and generally behaved like the Ukraine's grown-up version of Heidi' (Neill, 1995, p. 13). When it was revealed that Helen Demidenko was actually Helen Darville, daughter of English immigrants, with no authentic connections to the Ukraine, a storm of previously unspoken protests broke concerning the anti-Semitism of *The Hand that Signed the Paper.* Her text had remained the same; what had changed was that when the author was revealed as an Anglo Australian, the text was no longer held to be an example of the multiculturalism of contemporary Australia. Without the author's multicultural authenticity, the work's warrant for expressing unpopular views was withdrawn.

Less than a year later, a second scandal broke concerning the Aboriginality of the greatest living Australian Aboriginal writer, Mudrooroo (Laurie, 1996, p. 2). Author of a series of acclaimed novels concerning urban Aboriginal life, Mudrooroo was a full professor and head of indigenous studies at a prestigious university. In his case there was no suggestion that he had planned a hoax. Instead, according to his siblings, he was misinformed about his father's ethnicity. For many years he had lived the life of the Aboriginal underclass with which he had identified and about which he had written so powerfully: an impoverished family; split up by child protection agencies; life in an orphanage; early convictions for assault; imprisonment; and so on. Aboriginal leaders have nevertheless been critical of Mudrooroo's unrepentant response to revelations about his ethnicity, calling for 'Aboriginal people to reclaim ownership of Aboriginal culture and to "out" pretenders' (Dixon, 1996, p. 6). Although a flood of outings identified a series of other assumed identities, including a non-Aboriginal man who assumed the name of an Aboriginal woman to win another literary prize, an eminent non-Aboriginal woman artist who admitted selling paintings she represented as having been painted by an Aboriginal man, and a man who assumed the identify of his nephew in order to win a youth literary prize, the news coverage of successive outings was increasingly muted. What remains, however, is a sense that even in fiction the authenticity of the text is underwritten by the reader's knowledge of the facts behind the narrator's voice. As Barone (1995) has said, 'the weight of the *Zeitgeist* . . . leans against writers who presume to tell stories of other kinds of people' (1995, p. 69).

In the context of writing about teaching, the same question of authenticity occurs when researchers co-opt the voices of teachers and include them in the researchers' narratives and arguments. Too often, we think, researchers visit classrooms looking for data to fit their own predetermined category systems. As they select, summarise and juxtapose the data, the teacher in whose class the researcher is a guest turns from a participant in the project — capable of engaging with the researcher and of expressing his or her own intentions — into a powerless object of the academic author's gaze. The participants' voices are drowned out by the researcher's voice (Clifford, 1988; Geertz, 1988), especially when the data are used in short decontextualised quotations selected to illustrate the author's argument. We think that the test of authenticity to be applied is a phenomenological test: the parts of an ethnographic narrative that apply to a participant should be recognisable to the participant and reflect their own language and construction of reality. In this sense, in Trinh's words, 'There must not be any lies' (1989, p. 143).

A second issue of voice and authenticity concerns the voice we have adopted in this book. The book is primarily a piece by two hands, but there are other voices represented. We, John and Bill, have worked together for more than a decade on these and other issues. We have come to see in similar ways the phenomena in which we are interested, and to favour the same range of interpretive strategies and kinds of conclusions. Although various pieces of text were first prepared by one or other of us, we occupy a similar interpretive space. Our collaborative work has created a territory of shared meaning, a voice which transcends the details of which one of us was the seeing 'I' when the data were collected, and submerges the seeing 'I' into an interpretive 'we'. We have used the same theoretical resources and empirical methods, and have long since become used to seeing pieces of each other's original text emerge in a text first drafted by the other author. The convention we have followed in this text has been to restrict the use of 'I' to the author who collected the empirical data, and to use 'we' to represent shared interpretations. In Chapters 2 and 7 the 'I' is Bill, who appears as a participant in the events in these chapters and also in Chapter 8. In all three of these chapters, a second voice is Johanna, a teacher colleague of Bill's who is quoted at length in her own words, and appears as a participant in Bill's narrative. John is the 'I' in Chapters 5 and 9, and John appears as a participant in the events in Chapter 8. Chapter 6 is written in the third person where the 'we' are both John and Bill, and the first person character is Gerald, a colleague who writes about his own learning as a teacher. Chapters 3 and 4 are written in the third person, and the 'we' are John and our colleague Helen Wildy, who is the joint author of these two chapters.

THE NEVER ENDING STORY OF INTERPRETATION

Although there 'must not be any lies', fidelity to the perceptions of the people in the stories does not close down the problem of interpretation. In writing ethnographic narratives, we are on a journey away from the data — what Latour (1987, p. 241) called 'a cascade of representation' — as we move from the experience itself, to representation as transcripts or field notes, to incorporation into an argument, and finally to creation of a text that submits to the rhetorical and theoretical structures of the field. At each of these steps a further layer of interpretation is produced by the authors. Working in the shadow of the Enlightenment, building logical structures and citing authorities in APA (American Psychological Association) style, the pressure is to close down the possibility of alternative interpretations.

Recognising that understanding is 'always more than the mere recreation of someone else's meaning' (Gadamer, 1975, p. 338), we have tried to create a text that remains open to interpretation. Principally, this has been done by building our interpretations on extended narrative descriptions of events, and by separating the narrative data from more analytical interpretations of events. No interpretations, however, can exhaust the possibilities for meaning in a story; within every narrative there are the resources for other stories which might have been told. As the novelist Michael Ondaatje has said, 'Never again will a single story be told as though it were the only one' (Ondaatje, 1987). Understanding, consequently, remains elusive. Despite our attempts to provide accounts based firmly on actual events, contextualised by the history of the participants and our own preconceptions, texts are always open to further interpretation. It is finally a matter of taste, we think, where to locate narrative research between the Scylla of under-interpretation and the Charybdis of over-interpretation. To make too little of the data, to expect that it can speak for itself, is as much to be feared as the overdetermined interpretations which are the Enlightenment legacy in science education. We have used several strategies to keep open the text to alternative interpretations. In Chapter 6, for example, we have located Gerald's story in the context of several other interpretations, before offering our own interpretation. In Chapter 7 and Chapter 9 we have used an alternative strategy, explicitly separating vignettes from the interpretations of the vignettes.

THE PROBLEM OF GENERALISATION

The goal of rationalist generalisation from one data set or case to another is an example of what Holmwood (1996) has called 'the fantasy of coherence', an appeal to a world that is less messy and more rational than the postmodern one in which we find ourselves. Faced with the inability to meet quantitative (positivist) standards for generalisation, some writers have distinguished between 'naturalistic generalisation' and 'scientific generalisation' (for example, Stake, 1988, p. 260). We think that this distinction is unnecessarily coy.

On the one hand, we are sometimes surprised at survey researchers' willingness to generalise — across cultures, for example, about attitudes to cheating or preferred learning environment — on the basis of brief translated survey instruments (Wallace & Chou, 1998). On the other hand, we doubt that researchers in the narrative/case study/qualitative traditions really believe that their conclusions are restricted to the cases which they have documented closely. Although their texts might carefully specify the limits to generalisation from cases, loose talk around corridors and conferences finds many researchers willing to generalise well beyond case boundaries.

Without generalisations beyond cases, few of the commissioned case study research projects we have undertaken would ever have been commissioned. Indeed, as we prepare the results and executive summary sections of such reports, we often imagine getting the same response to our thick descriptions as Edward Gibbon got from his patron, when he presented him with Volume II of *The Decline and Fall of the Roman Empire*: the Duke of Gloucester is alleged to have said 'Another damned thick, square book! Always scribble, scribble, scribble! Eh, Mr Gibbon?' What we imagine our sponsors saying is: 'So you've collected all this detail about how a few physics teachers implemented the new syllabus, but what does it mean in general, what should we (they) be doing differently in the state-wide implementation project?' While the fiction is maintained that researchers do not generalise beyond cases, these are hard questions to answer. Rather than refusing to generalise, we prefer to pay attention in a disciplined way to which anecdotes are collected in case studies and then to make claims about the representativeness of those anecdotes. As Greenblatt has written, in the context of travellers' tales:

If anecdotes are registers of the singularity of the contingent . . . they are at the same time recorded as *representative* anecdotes, that is, as significant in terms of a larger progress or pattern that is the proper subject of a history . . . A purely local knowledge, an absolutely singular unrepeatable, unique experience or observation, is neither desirable nor possible, for the traveler's discourse is meant to be useful . . . Anecdotes are among the principal products of a culture's representational technology, mediators between the undifferentiated succession of local moments. (emphasis in original) (Greenblatt, 1991, p. 3)

What follows, we hope, is more than a succession of local moments. In Section II of the book, *Teaching*, we examine the practice of four teachers with whom we have worked. Chapter 2 provides an account of the relationship between knowing and teaching, using narrative data drawn from Bill's collaborative work with Johanna, a teacher of grades 7 and 8 in a small Canadian junior high school. Using descriptions of Johanna's settled teaching in familiar and unfamiliar contexts, the chapter argues that knowledge of teaching is essentially historical, shaped by teachers' biography and experience[2], and encoded in a settled repertoire of teaching. Chapters 3 and 4 expand this argument, using narrative data drawn from John's account (jointly authored with Helen Wildy) of two Australian grade 11 physics teachers, as they grapple with the implications of a new, constructivist physics syllabus. Through an analysis of the influence of biography, tradition and experimentation on the changes made by David (Chapter 3) and Mr Ward (Chapter 4), these chapters argue that the changes teachers make are the result of interpretations of the present which can only be made with reference to their understandings of the past.

Chapter 5 fills out this account with some alternative perspectives on teachers' learning. It describes the work of a teacher we call Ms Horton, an experienced biology major teaching the topic Chemical Change for the first time. Unlike Chapters 2, 3 and 4, which focus on the teacher's perspective, Chapter 5 provides a narrative description of a lesson sequence followed by four alternative interpretations based on interviews with Ms Horton, her head of department and two students. The chapter concludes that there is more to understanding Ms Horton's teaching than she is able to tell. In addition, it is necessary to understand the alternative realities of students such as Karl, who is only loosely connected to Ms Horton's goals for teaching; Punipa, a keen and successful student who was frustrated by Ms Horton's open-ended tasks; and Mr Greg, Ms Horton's head of department, who was anxious about her capacity to cover all of the content he felt was essential.

Section III of the book is about teacher reflection. Chapter 6 examines the potential of teacher-written narrative cases in helping teachers to reflect on their practice. It provides a narrative account prepared by a grade 10 teacher with a biology major who was working outside his area of specialisation. The chapter includes commentaries by the teacher, Gerald, by his head of department and by a student teacher, and discusses the way in which such alternative commentaries can be used in teachers' pursuit of learning about their teaching. Drawing again on interviews and observations from Johanna's classroom, Chapter 7 provides an account of the ways in which reflection contributes to teachers' learning. We characterise reflection in terms of two dimensions, *interests* and *forms* of reflection, building in different ways on the theoretical work of Schön (1983, 1987) and Habermas (1971).

Section IV of the book, *Collaboration*, takes up two substantive issues about research on teachers' learning: working together (Chapter 8) and ethics (Chapter 9). Chapter 8 uses narratives about two pairs of teachers (Johanna and Bill, and Geoff and Amanda) to identify the qualities of collaborative partnerships: similarities and differences, symmetry, trust, emergence, humility and fair exchange. Chapter 9 uses four short narratives to identify a series of lessons we have learned about conducting ethical narrative research.

Section V or Chapter 10 is titled *Teachers' Learning and the Possibility of Change*. In this final chapter, we draw together the arguments made in the preceding chapters into a set of propositions about teachers' learning.

NOTES

1. Carter (1993), Clandinin and Connelly (1996), and Polkinghorne (1995) provide extended discussions about the use of stories in research on teaching. Phillips (1997) offers a critique of several of these accounts.
2. Throughout the book, we use the singular convention for terms such as teachers' experience, biography, history, knowledge and understanding while acknowledging that one teacher's experience etc. differs from another.

SECTION II

TEACHING

2. KNOWING AND TEACHING: JOHANNA

This chapter explores the relationship between knowing and teaching, using examples drawn from a year-long collaborative research program in one teacher's classes.[1] The specifics of the chapter concern my collaboration with a teacher we called Johanna (Louden, 1991). She and I worked together teaching art, music, writing and science to several groups of grade 7 and 8 students in a small junior high school in a large North American city. In the time we worked together our roles as teacher and researcher varied. At the beginning I worked as an observer of her lessons. The longer I stayed in her room, however, the more active I became as a participant. Eventually, I found myself working as a teacher, while Johanna observed and critiqued my work.

The narrative which follows is organised essentially in the same order as the events occurred. The account begins with an overview of some of the tried and true lessons in Johanna's settled repertoire of teaching. In the first section I draw attention to the importance of patterns of teaching, the consequences of teaching familiar content, and the impact of a teacher's biography and experience on her teaching. The second section of the chapter describes a series of lessons we taught together, when Johanna — an arts specialist — decided that she wanted to learn more about how I taught English. In the account of our jointly constructed writing lessons, I draw attention to a dilemma that frequently arises when teachers try to learn new ways of teaching and describe the process by which Johanna assimilated some new content into her existing patterns of teaching.[2] The third section of the chapter follows us through a similar process as we tried to learn some new ways of teaching in an unfamiliar teaching area — science — drawing attention to the impact of our old patterns of teaching, our poor content knowledge and our imperfect resolution of the dilemma of education versus classroom control in this new context.

Bringing these ideas together at the end of the chapter I offer an interpretation of the ways knowledge of teaching grows and changes, emphasising the importance of teachers' preconceptions in shaping their understanding of what it means to teach, and conceptualising growth in teachers' knowledge as a gradual reshaping of preconceptions as teachers encounter and overcome new gaps in their understanding of teaching.

These ideas about teachers' knowledge emerged in the context of several vigorous programs of research on teachers' knowledge. They draw directly on what Fenstermacher (1994, p. 13) has called the Elbaz/ Clandinin/Connelly strand[3] and the Schön/Russell/Munby strand[4] of research on teachers' knowledge. Like Elbaz (1983), my interest has been in how teachers' personal knowledge reflects their history, context and intentions; like Connelly and Clandinin (1985, 1990) I have come to see stories about teaching as 'the closest we can come to experience' in the study of teaching (Clandinin & Connelly, 1996, p. 29). Like researchers in the Schön/Russell/Munby strand of research, I have been especially interested in the knowledge teachers draw from their experience; the 'swampy lowland' of practice rather than the 'high, hard ground' of technical rationality (Schön, 1983, p. 42).

KNOWING AND TEACHING: JOHANNA'S REPERTOIRE

Patterns of Teaching

Some of Johanna's patterns of teaching were fixed and present on almost every occasion I saw her teach. She usually began class discussions by relating some personal experience of her own, and she usually stood at the door at the end of a lesson and asked people whether their table captain had said that their table was tidy enough for them to leave the room. Other patterns had been developed to match the needs of particular subjects, such as her system of distributing guitars for whole class music lessons and of singing over the students as they picked out the chords. Another pattern of teaching concerned her use of a blue oriental rug in the centre of the art room as the physical focus of activities: a place to display completed art work, to perform music appreciation activities, or to gather for a class meeting.

These patterns of teaching allowed Johanna to resolve familiar classroom problems in ways that were predictable and comfortable for her. When she conducted class discussions, for example, she worked to involve everyone in the discussion by asking people to gather in closely together on the blue rug, by ensuring that no one sat outside the circle, and by insisting that she was able to see every face. She helped people to make contributions to the discussion by using small groups to ensure that people were not too intimidated to speak in the large group, by working her way around the circle when she was sure that everyone had something to contribute, and by acknowledging but not judging the comments students made.

This class discussion pattern, and other patterns of teaching she used, provided Johanna with ways of organising the class which could be matched to a variety of content. In one art lesson, Johanna used the class discussion pattern to explain the concept of value (light and shade) in graphics. In another she used the class discussion pattern to help people understand a challenging visit to an art gallery. In a third case Johanna used the class discussion pattern to help students learn from the experience of camping out, being away from home, and having to organise and prepare food together in groups.

Johanna used a wide variety of sources for her teaching content. Some of the content, concepts in art theory and music appreciation, for example, came from the parent academic disciplines. Sometimes she drew on her knowledge of art theory to help students make more of a current school activity, such as the displays prepared following a visit to a farmers' market. On the other hand, the content for a discussion following an art gallery visit came from Johanna's current artistic interests and her belief in the importance of people learning to appreciate the beauty around them. In other cases, Johanna paid little attention to the formal content of the curriculum, preferring to pursue learning goals which were central to her hopes and dreams for education, helping students to become independent and responsible adults. Her judgements about what content was relevant and useful to teach, then, were neither shaped nor constrained by school board guidelines in the subjects she taught. Instead, she matched the needs and interests she saw around her with content and patterns of teaching she carried forward from year to year.

In Johanna's teaching, her past, present and the future ran together. In order to fully understand the ethnographic present of her teaching it is necessary to look both behind and beyond the surface patterns: looking back towards the history which shaped her teaching, and forward to the predispositions to future action they represent. Describing the present alone is insufficient. As the hermeneutical philosopher Gadamer (1975, p. 321) put it:

'To recognize what is' does not mean recognizing what is just at this moment there, but to have insight into the limitations within which the future is open to expectation and planning.

Biography and Experience

What was 'just at this moment there' as I observed Johanna teaching was a series of patterns of teaching and some particular pieces of educational content. Behind these patterns of teaching and their educational content stands Johanna's biography and experience. The patterns are not arbitrary; they represent Johanna's best solutions to the familiar practical problems she faced in the classroom. To take the example of her whole class guitar teaching, the pattern of teaching she used solved an elaborate set of practical problems in a way which was consistent with her sense of herself as a teacher in general and a music teacher in particular. Johanna once told me that giving a full class of adolescents a guitar is a bit like giving them each a machine gun: it is inevitable that they are going to wave them about, play with them, pretend to know how to use them, and see what sort of noises they can make. The trick for Johanna had been to find a way of them all having a guitar without anyone getting shot. She did not want to have to be authoritarian about the need to be silent when she was giving an explanation, nor did she want to have to badger and remind them, but she believed that she must have complete silence while she was tuning, demonstrating or making a teaching point. The way in which she did this was learned many years before I met her, when she taught music full-time, repeating each lesson as many as seven or eight times a week. After years of practice, Johanna had now refined the whole-class guitar lesson to the point that she was confident that she could handle any of the problems that were likely to emerge. As she said:

I have learned how to do a guitar lesson that works. All the possible behaviours that kids can throw at me around guitar have happened, I have dealt with them, and now it happens successfully. . . . Almost always, because there's always one kid who will surprise you. Almost always I can account for what is going to happen. That's where the experience counts. I'm comfortable and I do a good job.

Her established patterns of teaching reduced the number of problems she had to deal with, and they did so in a way that was consistent with her sense of herself as a teacher. Her limited formal training in music, combined with a belief that her task was to introduce all students to the pleasure of music making, meant that she valued the confidence and participation of many students more than the musical excellence of a few students. She knew, from her own experience as an adult learner of the clarinet, how easily beginning musicians may be discouraged from participation and she was dedicated above all else to ensuring that no one became discouraged from making music by her guitar lessons.

When Johanna was working deftly on lessons in her settled repertoire she solved the practical problems of teaching so well that they became almost invisible. There was no sense that she had to work hard to motivate, control or discipline the students. Her classroom control was mostly invisible, submerged in the patterns of teaching she had developed: she told students where to sit for particular lessons, insisted on absolute attention during guitar lessons, carefully marshalled who would speak next in her class discussions, but rarely had any cause to show irritation or to raise her voice. In the most delicate of her interactions with students, in class meetings, her control was even more muted. Her actions were guided by the belief that learning is more likely within a safe and controlled environment. As she said:

Part of [my] understanding of learning is that it can only happen in a situation where there is agreement to participate in it. In a situation where there is no control, where the students are not exhibiting any self-control, there's no agreement to go ahead with learning.

KNOWING AND TEACHING: WRITING

When Johanna and I decided to collaborate we had no particular plans for me to act as a teacher in her class; our initial understanding was that I would watch and she would teach. We stumbled into teaching writing together because Johanna was dissatisfied with the quality of some student writing she had planned to use as text for an illustrated book project in art. She tried to improve students' writing by teaching a formal lesson on punctuation, and at the end of the lesson felt a little self critical. She asked me what I thought and when I responded, I focused on issues of control: chiefly whether the students knew what to do and seemed satisfied with their own work. My guess was that they found the task of correcting punctuation on a handout familiar, obviously legitimate for a teacher to set, and — best of all — easy to do and get right. Dissatisfied with my response, Johanna asked, 'Do you think they learned anything?' Here she was grasping one of the horns of what seems to me to be an essential dilemma in teaching: the issue of education versus social control (Sharp & Green, 1975). On one hand, her lesson seemed perfectly satisfactory to the students; it matched their sense of a properly organised lesson. Johanna, however, would rather have traded some of the control in the lesson for more serious educational outcomes.

Johanna asked me to describe some other ways of teaching the same content, and this led to a long series of discussions about how to balance the demands of education and classroom control in teaching writing. Over several weeks, we worked together trying to achieve Johanna's educational goals without compromising her expectations about classroom control. At the end of the time we spent together, Johanna had been able to assimilate some of my craft knowledge into her established patterns of teaching, especially where my process-based approach to teaching writing was similar to her pattern of over-the-shoulder assistance in art lessons. When this worked well she was able to achieve the kind of invisible classroom control I had seen in lessons involving more familiar content. The tension between activities which maximised control and activities which maximised her educational goals was reduced and she was able to resolve the new set of education-control dilemmas which had been introduced by her decision to pay more attention to a familiar content area.

Horizons of Understanding

Johanna's new patterns of teaching writing and their resolutions to new forms of the education-control dilemma were shaped by the predispositions to action which were carried forward from her previous experience of teaching and by her biographies. Her writing teaching involved a subtle layering of new knowledge over old knowledge of teaching. One way of conceptualising this process is to use a concept drawn from philosophical hermeneutics. Describing the process by which the understanding we bring from the past is tested in encounters with the present and forms the understanding we take into the future, Gadamer (1975, p. 273) uses the term *fusion of horizons:*

> The horizon of the present is continually being formed, in that we have continually to test all of our prejudices. An important part of this testing is the encounter with the past and the understanding of the tradition from which we come. . . . In a tradition [the] process of fusion is continually going on, for there the old and the new continually grow together to make something of living value, without either being explicitly distinguished from the other.

Transposing Gadamer's terms from the hermeneutic understanding of written texts he had in mind to the understanding of action which is my concern, we may see the *horizon* as the predisposition to act in certain ways in the classroom.[5] Johanna's predisposition to act, I believe, was shaped by her understanding of her own biography, and by the patterns, content and resolutions to the education-control dilemma which together make up her repertoire.

For Johanna, one of the key tests of an appropriate program was that we found a way to help students be more productive in their writing without discouraging less able students. As she said:

> It's the whole crux of the issue of [this school]. At what level can you let them go? You cannot let them become discouraged, to give them time to waste. They can't handle it. [One] way was to let them fail miserably and let them suffer the consequences and then realise that they would have to pull their socks up. My feeling about that was that it was such a discouraging experience that it made them angry and they ended up doing it for the teacher, not for themselves.
>
> I think with [the grade 8 class] we are going to have to give them a lot more help. With most of those kids last year the only thing that really helped them was a personal conversation telling them what they had to do. Last year, whenever I had a project due I would write up all the steps. They couldn't read it if I handed it out.

Because this was Johanna's understanding of the problem, we gave the students 'a lot more help' to complete their writing, in a way that would not involve individual discouragement, and which used some of Johanna's well-practised patterns of art teaching and some of my patterns of writing teaching. She was able to merge the old with the new, gradually developing new horizons of understanding which included some new teaching content and processes.

KNOWING AND TEACHING: SCIENCE

Not all changes in teaching are as successful and satisfying as our experience of teaching writing together. Johanna and I had a much less satisfying experience in trying to find a comfortable way to teach science, a subject which was unfamiliar at the beginning and still a mystery to us both at the end of our collaboration.

Johanna inherited responsibility for the school's science program as a result of the reduction in the school staff from four to three teachers. The science specialist moved on and Johanna was obliged to take on a whole new subject area. Creatively, she decided to combine her need to learn to teach science with an offer I had made to help her in any way I could. She asked me to teach the grade 7 and 8 science program. As a former secondary English teacher whose science education had ended with grade 12 chemistry almost 20 years before, this was not an area of strength for me but I was willing to have a try.

Teaching Unfamiliar Content

Unfamiliar with the content, my first strategy was to locate the school board curriculum materials. I prepared some very conventional practical lessons on the first topic, mixtures and solutions. After many years of English teaching, I felt like a beginner again, failing to solve simple practical problems which I would not have encountered in an English class. For instance, in one lesson I had been working hard to draw together the result of the lesson into an authoritative set of blackboard notes which the students were intended to copy into their work books, and I was unaware that only half the students had succeeded in completing and copying down the summary chart of the characteristics of a solution.

More generally, I also began to realise how large an impact the preparation of demonstrations and experiments made to my patterns of teaching. Used to the limited physical preparation of English teaching, and in the absence of equipment which could be damaged or dangerous, I had learned to expect that I could trust my instincts and spontaneity as lessons developed. In science, however, I found that the commitment I had made to physical preparation and the need to have all of the equipment packed away by the end of the lesson led me to follow my lesson plans much more carefully.

I was quite pleased with these rather conventional lessons. The students seemed to be doing some science, I didn't make too many blunders with my limited content knowledge, and we were able to get the equipment out, used and packed away in time. Johanna was more sceptical about the value of these lessons. She didn't mind when I broke away from formality and gathered the students around the bench for a demonstration lesson, but none of the content I was teaching about mixtures and solutions seemed very interesting to her. She said that she was sure that it was the same for the students. I was cautious about changing my approach: at least it was safe and predictable. But Johanna was unconvinced. As she said:

[To] teach science you have to find out what they are really interested in, what they would be willing to do. You can even try and make them interested in solutions, and you can stand on your head and spit wooden nickels — which is basically what we have been doing — and bamboozle them into being interested in solutions. Or, you can do less work, or the same amount of work, with something that really gets them going.

The kind of school science I was able to offer was in direct conflict with her sense of good teaching. As she explained:

I know what I want to do, I know what I want it to look like, what I want to have happen. I want to have a class where the kids are paying as much attention as they were [in a recent nutrition class]. They were all asking questions about these diseases that either killed their relatives or thought might kill them. I want them to want the information that they have to look for. So that they are actually doing scientific research and learning about science because it is something that concerns them. Boy, they are really interested in health. They are interested in the ways their bodies work and they are interested in sex, so it seems that we could do that and call that biology and no one's going to complain.

Johanna suggested that science for the rest of the term should be based on a long biology assignment dealing with the sort of health and disease issues which we knew interested our students. They could work in pairs on different topics and then make a presentation to the class at the end of the term. I gave up my attempts to 'bamboozle them into being interested', and went in search of books that could support a research assignment.

Instead of developing the new patterns of teaching that were implied in the school board science syllabus, we tried to fit the new content of science inside our old patterns of teaching. When we did this, we each contributed something to the task from our existing patterns of teaching. For example, when I planned to explain to the students how to write up an experiment as their new assignment required I found myself abandoning the lesson I had planned and spontaneously using a game show format I had often used as an English teacher. Following my instructions the students demonstrated an 'experiment' in the personas of the host and assistant of 'Wheel of Fortune'. It didn't feel much like my outsider's sense of a science lesson, but the lesson worked well because I was working within an established pattern of teaching learned in the more theatrical context of my work as an English teacher. Similarly, Johanna contributed to our science teaching the pattern of independent research assignments which I had seen her use in a series of health and nutrition lessons.

The Dilemma of Education and Control, Revisited

Throughout our science program, Johanna and I faced a series of education-control dilemmas. Whereas I was pleased with the first few science lessons because they met my beginners' control requirements, Johanna had reservations about their educational value. Equally, from the first time we had talked about the prospect of teaching science she had expressed doubts about any of the school science programs she had seen. Instead of trying too hard to make the connection between the syllabus and the students, her preference was to allow students to work with content that they already found relevant and to conduct their own research. When she suggested that we use assignments to be completed at home, one of her established patterns of teaching, I had both educational and control reservations about her plan. Would they be able to find the appropriate material, I wondered, and would the grade 8 class be sufficiently independent to complete such an assignment? She acknowledged that it would be 'tricky' to control but thought that the assignment I prepared had more merit than continuing to 'bamboozle them into being interested'.

Johanna and I did not always agree about the resolution of particular education-control dilemmas. In the case of the game show lesson, for example, Johanna enjoyed the lesson and thought that it was a success but I thought that the educational content was relatively thin. In part this reflected the connection between the lesson and my own personal interests. Just as Johanna was enthusiastic about lessons which might help people make decisions about which research they ought to trust, I was interested in lessons which explored the relationship between theory and a body of evidence.

By using familiar patterns of teaching, we were able to submerge the new control problems that arose in teaching an unfamiliar subject. Recycling my English teacher's patterns of teaching, I no longer had to prepare step-by-step lesson plans and did not have to master the complexities of practical lessons. When our science lessons ended, I was still unsure about whether they had been a success. The control problems had disappeared, but the educational value of what we had done was still unclear to me. Unlike the experience of teaching writing, our horizons of understanding were not stretched by the gaps in knowledge we confronted in teaching science. Instead we resolved the new education-control dilemmas we faced by retreating to well established patterns of teaching we had developed in the context of more familiar content.

CONCLUSION

Over time, Johanna developed an extensive repertoire of teaching: a set of standard patterns teaching, familiar content and effective resolutions to common pedagogical problems. This repertoire involves a set of practical responses to a basic education-control dilemma: how can she help individual students learn in an environment which requires her to control the behaviour of the group? For a skilful and experienced teacher, such as Johanna, the education-control trade-off is almost invisible. She appears to be able to achieve order and group cohesion without dampening the possibilities for individual learning. These patterns, content and resolutions to the problem of education and control are not arbitrary, but are historically based in Johanna's biography and experience as a teacher. Together, her repertoire forms a predisposition to act in the future, what has been called her horizon of understanding.

This horizon is not static but is constantly in the process of formation. Confronted by new problems in teaching, Johanna struggles towards a fusion of horizons. She attempts to solve new problems in ways that are consistent with the understanding she brings to the problem, a process which leads in turn to new horizons of understanding about teaching. These new problems expose the education-control dilemma, and Johanna works to include new resolutions in her repertoire and to drive the education-control dilemma back underground. For some problems, her current repertoire may gracefully be extended to include solutions for the new problems; in other cases this is more difficult, and uncomfortable trade-offs between education and control are necessary.

When the gaps in understanding Johanna encountered in the illustrated book project were relatively small, she was able easily to add new patterns of teaching to her repertoire. These patterns, based on a view of writing teaching which I had previously used, were consistent with her hopes and dreams for education. The new content knowledge she needed to teach writing was consistent with the craft knowledge she already had about art and music, and was readily available from me. And furthermore, the practice of making time for writing conferences with students allowed Johanna to find a new context for the problem-solving discussions which are so central to her teaching.

When we tried to teach science, the gaps in understanding Johanna and I both faced were much larger. In order to take on the 'school science' of the guidelines, Johanna would have needed to learn new subject content, and would have had to believe more in the value of science. As it was, she was rather too sceptical to make the effort to bridge the gap between her horizons of understanding and the notion of teaching implicit in the guidelines. Instead, we both bridged the gaps in our understanding by changing school science into something more familiar to us. For me, this meant using some of the patterns of teaching I carried forward from my English teaching, and for Johanna this meant using science as an opportunity to pursue her larger goal of helping students become independent learners.

This account of change in what Johanna knows and teaches has taken an hermeneutical view of teachers' knowledge. Rather than focusing on Johanna's knowledge as a static body of knowledge, I have argued that the essential characteristic of her knowledge is that it comprises layer after layer of interpretations, laid down over many years of teaching. These interpretations are encoded and visible in the patterns of teaching and resolutions to education/control dilemmas which comprise her repertoire. Her understanding of teaching is essentially historical, shaped by her biography and experience and the preconceptions she carries forward to each act of interpretation (Gadamer, 1975). Her understanding of teaching, like mine, is also incomplete. It continues to grow and change with each encounter with new content and teaching contexts, with each gap in understanding she faces in her work as a teacher.

NOTES

1. This chapter contains reworked material which first appeared in a 1991 book by William Louden titled *Understanding Teaching: Continuity and Change in Teachers' Knowledge* (London: Cassell & New York: Teachers College Press). It is reproduced here with the permission of the publishers.
2. Berlack and Berlack (1981, pp. 135–165) and Tripp (1993, pp. 49–51) have used the term 'dilemma' to describe tensions teachers face in choosing between mutually exclusive actions. My use of the term follows Lampert (1984) and Olson (1985), both of whom speak of dilemmas in terms of the practical problems which must be resolved by teachers in action.
3. Clandinin and Connelly (1992) provide an overview of their program of research in the context of a review of the literature on teachers and the curriculum, and Clandinin and Connelly (1997) provide an account of their narrative research methods. Clandinin and Connelly (1996) provide a response to Fenstermacher's critique of the 'Elbaz/Clandinin/Connelly strand'.
4. Fenstermacher identifies edited collections by Russell and Munby (1992) and Clift, Houston and Pugach (1990) as examples of research in this tradition. Munby and Russell (1995) provide a response to Fenstermacher's critique of the 'Schön/Russell/Munby strand'.
5. Gadamer's term 'horizon' had previously been used by Nietzsche and Husserl to describe the finite understanding available to a person at a point in space and time.

3. TRADITION AND CHANGE: DAVID

Johanna's story demonstrates that the complexity of science teachers' work and knowledge often goes unrecognised because the complexity is largely invisible. Beneath the smooth surface of any experienced teacher's practice lies a set of tacit and explicit knowledge about what is worth teaching, about students' needs and abilities, and about the context in which they teach. To understand how and why teachers follow their patterns of practice we must see their actions as part of a much larger body of knowledge, reaching back into the past and creating purposes for the future.[1] The meaning of these patterns only becomes clear when set in the context of a teacher's personal and professional history and hopes and dreams for teaching. This chapter employs narrative as a way of representing and understanding the work of a second teacher — David — as he attempted to come to terms with the meaning of syllabus change. It examines the strategies of this teacher as he coped with the challenges of teaching physics in a new syllabus context and, in doing so, attempts to understand how teachers' knowledge is derived, how it is used and how it changes over time.

The data for this chapter were collected in David's physics classroom over a ten-week period. Throughout this time we[2] observed David at work and took part in his lessons. Informal discussions were conducted with David and his students during and after the lessons to gain insight into their understandings of the experiences. Our aim was to create a jointly acknowledged and mutually beneficial partnership with David — what in Chapter 8 we call 'fair exchange'. In return for allowing us into his classroom, David called on our collegial support and used us as a sounding board for his ideas. The vignette or story of David's teaching which follows was constructed by us from field notes and transcripts of interviews. Following the story is our interpretation of the events of the story and of the wider context of David's practice. The story is a partial view of David's teaching selected by us as researchers to illustrate aspects of his work and to highlight our broader understandings of how teachers' knowledge is formed and used. These understandings both shape and are refined by the construction and interpretation of the story.[3] However, with each new reading of the text, interpretations inevitably grow and change. In succeeding chapters, we will revisit our interpretations of David's work in the light of data from the classrooms of two other teachers.

The study was conducted in David's grade 11 physics classroom during the implementation of a new state-wide physics syllabus for students at the senior level of high school (Parker, Wildy, Wallace, & Rennie, 1994). The new syllabus signalled a marked shift away from a traditional algorithmic approach towards an emphasis on language and communication, qualitative explanation, problem solving, practical application and experimental work. Inherent in the philosophy of the new syllabus was an expectation that teachers would adopt more student-centred teaching strategies than had traditionally been the case. David, one of the most senior physics teachers in the state, had been involved in the discussions preceding the development of the new syllabus. He had also been selected as a 'link' or 'lighthouse' teacher to introduce the new syllabus to teachers in neighbouring schools. The study took place in the first year of the new syllabus while David was developing his understanding of the philosophy and strategies inherent in the new approach.

TEACHING PHYSICS: A VIGNETTE

The focus of David's teaching during the period of the observations was the new syllabus topic of Light and Sight. This vignette describes the sixth lesson on the topic.

The students were seated and talking quietly when David arrived for the lesson. He began the lesson by asking questions arising from the previous lesson on Snell's Law. He asked about students' understandings of concepts such as refraction, angle of incidence and angle of refraction. After a few responses David introduced the task for the lesson: to trace the reverse pathway of light through a semi-circular glass slab:

I won't tell you how to set it up. You'll need to talk about it between you. You work out what you need to record to answer the questions on the board:

1. What happens as the angle of incidence increases?
2. What is the ratio sin i/sin r? How does this compare with your result from the last lesson and what you know about Snell's Law? Explain this in your own words.
3. What does Snell's Law reduce to when the refracted ray disappears?

The way the students organised themselves followed a similar pattern in each lesson observed: the students sitting at each bench formed a work group and, after a few exchanges, stood up and walked to the equipment cupboards at the side of the room to collect whatever equipment was needed. In this instance they assembled slabs, stands, lights and power boards which they set up in a practised manner on their benches. There was little discussion and very quickly rays of light were being cast through the slab at its curved edge. Several groups did not understand that the light needed to be aimed at the centre of the semi-circular slab to make an angle with the normal. All groups drew and measured angles of incidence and angles of refraction, setting out their data in tabular form.

Two students, Craig and Diem, explored the angles without measuring them and quickly, and excitedly, found there was a point when refraction stops and all the light was reflected. They then went back to measure the angles carefully and record the results: they worked with concentration, plotting, calculating, graphing and writing.

After 20 minutes David stopped the class and asked one of the students, Shane, to show what happened to the light as the angle of incidence increased using a drawing on the blackboard. Shane drew the path of the refracted light and then, with some coaxing from David, illustrated the point when it disappeared and turned in the other direction. David described this as light which was reflected and at that point there was total internal reflection. Suddenly a student, Michael, called out: 'But it's always there. It just gets lighter as you get further away from the normal'. David reinforced Michael's discovery: 'Yes, we do always get some reflected light. Think of a car window — you see through it but you also see yourself reflected. So it's a mistake to ignore it and think that light is only reflected at the point at which it is totally internally reflected'. David did not attempt an explanation of the physics behind Michael's discovery.

Students spent the last part of the lesson writing descriptions of their findings in their own words. David stressed that what they wrote should be understood by any reader — scientist or non-scientist. Most of the students used a traditional form of laboratory report to record their findings although some used a more narrative form. In the closing minutes of the session, students packed away the equipment as David asked that the questions set for this lesson be completed for the following lesson. Although it was now lunch time several students lingered to discuss what they had done during the lesson. It was evident that the experience of experimenting with the light rays and the glass slabs had brought a deeper understanding of the concepts of refraction and total internal reflection. In the words of one student, Diep, 'I just had to accept it before when I read it from the book. Now I can talk about it. You actually get to see how it works when you do it yourself'.

As in all the lessons observed, David spoke very little to the whole group. When the students were working on their tasks, he stood among the benches, occasionally watching what groups were doing, sometimes asking a question, sometimes commenting. He spoke quietly, usually smiling and making gentle jokes and understated remarks of encouragement. From time to time David explained his interpretation of the intention of the new syllabus. He wanted to 'encourage students to understand what they are doing and therefore enjoy it, and to encourage all students — those from different backgrounds — to study physics'.

This vignette illustrates many of the dilemmas faced by David as he grappled with the new syllabus. The lesson commenced in traditional fashion with David revising some of the concepts which were to form the basis for the practical work to follow. The practical itself was conducted in a practised manner and the outcome of the lesson quite predictable — to confirm the operation of Snell's Law. However, the lesson was quite different for David in some subtle ways. The practical task was framed for the students as a series of questions rather than as a set of instructions to be followed. David treated Michael's observation as an interesting and worthwhile discovery rather than an aberration. He encouraged students to use narrative forms of recording their results rather than insisting that everyone produce a formal report. David blended these 'new' strategies with his 'old' and tested ways of teaching physics. Experience told him that students needed a knowledge of the application of Snell's Law and that this must be obtained through formal practical work and teacher-led discussion. David's understanding of the new syllabus told him that students needed opportunities to experiment with ideas in different ways if they were to understand the physics behind the formula. These strategies were reinforced by comments made by some of his students at the end of the lesson. To gain a deeper understanding of the how and why of David's pedagogy, it is necessary to look behind and beyond the surface patterns of the vignette; look behind the story for the history which shaped it, and beyond the story for the messages it carries about teachers' knowledge.

UNDERSTANDING DAVID'S TEACHING

David was an experienced physics teacher confronting a new physics syllabus. The syllabus brought with it a mixture of old and new beliefs and attitudes about 'good' physics teaching. Carrying over from the old was a belief that physics was a systematic and rigorous means of describing the natural world in terms of principles and laws and that students needed a basic knowledge of the content of these principles and laws. The new approach aimed to contextualise and humanise the subject, focusing on students' understanding and application of physics concepts. Reconciling the old with the new became the challenge for David and other physics teachers; how to suspend the certainty of their formal knowledge of physics to allow for the uncertainties which flow from finding the physics in the context. Debates about how (and whether) this could be achieved had been played out in physics teacher forums and letters to the local press. Another significant issue was the influence of the external examination — a state-wide test used for determining university entrance. Many physics teachers had publicly expressed concerns about how the philosophy of the new syllabus would be incorporated into a traditional written examination format. It was within this context that David was trying to teach the lesson on Snell's Law, refraction and total internal reflection.

We have constructed this story of David's teaching with a particular purpose in mind, to highlight our theoretical concerns about teachers' learning. The story, and the meaning we attribute to the events in the story, flows from our reflections on our experiences in composing this and other stories (Schön, 1983). In analysing the story, we impose an order and meaningfulness that may not have been apparent in the event as it happened (Polkinghorne, 1997). In the hermeneutic tradition, we have created meaning from the text or experience in the light of our preconceptions, interests and research frame (Wallace & Louden, 1997). Thus, in keeping with the ideas introduced in the previous chapters, the themes we use in this chapter — change, tradition, experimentation and biography[4] — are our attempts to make sense of the connection between David's practice and the knowledge he has and uses.

Change

The new physics syllabus placed store on the importance of students' exploration and qualitative understanding of the subject matter. David was sympathetic with this view of teaching and learning. He wanted to enhance students' comprehension by providing opportunities for them to plan, try, make mistakes and talk about their work with peers. While the lesson on Snell's Law had many features of a traditional physics lab, it was also more open-ended than his usual practice of giving step-by-step instructions. In previous years, his instructions would have included information about the angle sizes to be measured and the form of the table into which the results were to be entered. In David's words, this was the 'traditional cookbook approach in which students relied on the teacher to prepare the experiment so that they would arrive at predetermined conclusions'.

In classroom activities and experiments David consciously tried to encourage students to 'have a go', 'do it on your own', and 'work it out yourself'. He commented during the lesson on Snell's Law, 'here's where I'm attempting to open it up a bit'. The unpredictability of this approach meant that David was uncertain about this lesson planning. He proceeded gradually and hesitantly, checking back over his progress and revising his plans as he went:

The interesting thing for me is that none of this is pre-planned. I'll go back and plan it after I've finished. I try things — I'm basically following through a bit of a sequence. I'll probably go back and do the planning afterwards.

While David had a generalised vision of where he was heading and the strategies he wanted to try, he allowed his lessons to evolve. However, when students arrived at unexpected outcomes, he was faced with the dilemma of how to deal with them. He wasn't sure of the significance of non-standard student answers and of their contribution to the scope and sequence of the physics course. On these issues he was at the limit of his pedagogical content knowledge (Shulman, 1987). David was much more comfortable when the outcomes were predicted and predetermined.

[In planning the experiment themselves] they don't necessarily get what you as teacher have in your mind that you want them to get. But they find some pretty interesting things. The question is: what do you do with what they find?

For example, in the Snell's Law lesson, David expected that students would discover the critical angle, the point at which all light rays are internally reflected; he did not expect them to find that a certain amount of light would always be reflected. Formerly, his teaching would have quickly closed down the discussion of this apparent anomaly. However, when Michael's explorations led to the announcement: 'But it's always there; it just gets lighter as you get further away from the normal', David created an opportunity to connect with students' everyday experiences by talking about reflections in car windows.

In this instance, David managed to deal with an unexpected outcome because of his physics knowledge base. However, in an earlier lesson, when students were dissecting bullocks' eyes, they asked questions about how the muscles adjusted the shape of the lens. David gave his interpretations of how they worked. Later, when two students offered a different explanation, David acknowledged that his knowledge was incomplete and that he too was learning. David mentioned this experience on several occasions and referred to it with his students frequently. He believed that, difficult though it might be, he needed to 'admit to students that he was not always right'. Modelling a questioning, challenging, inquiring attitude was part of the process of change for David.

Tradition

While recognising the need to teach for understanding and to place less emphasis than previously on content coverage, David struggled to re-form his own vision of 'good' physics teaching. He analysed what he was trying to do and achieve against his background of 26 years teaching physics.

> There are a lot of questions I ask myself all the time. . . . You've got to come to grips with what's the better way to go. I know I enjoy it when I do it this way. But society says schools are about knowledge, facts . . . The hard bit is to shift away from what society has always expected, from the emphasis on rote learning of content.

While David referred to certain external societal pressures to conform, he acknowledged that these were reflected in his own traditional values in relation to the changes.

> I'm fairly traditional so I'm looking for the content they have learned rather than how much can they do, what they understand, how much they enjoy physics.

At the start of the first lesson on Light and Sight, for example, he explained to his class that he was trying out a new way of teaching physics and that he was unsure of its outcomes: 'I'm experimenting. I'm not sure how this'll be for you'. David was less certain of his commitment to the values underlying the new approach to physics teaching. He was troubled by some nagging doubts about the consequences of the 'context first' approach — doubts about his ability to teach in this way and doubts about the impact of the external examination.

> I often wonder whether I'm doing the right thing. It's hard for me because I honestly don't know if I really believe it myself. . . . I'm still a little uneasy, probably because I'm conservative . . . And we still don't know what will happen with the [external examination].

The tension between meeting content objectives and teaching for understanding caused continuing concern for David. In the lesson described earlier, he was clearly concerned that his students had a sound grasp of the nuances of Snell's Law. In many respects the practised manner in which the students set up the equipment for the lesson was similar to the prescribed laboratory sessions of the old syllabus. It was only later, during the class discussion on total internal reflection, that we saw some evidence that David was prepared to allow the lesson to diverge from the norm. He encouraged Michael's non-standard observation and used a contextual example to illustrate the point. Even though David was comfortable with this amount of experimentation with the teaching of the Sight and Light topic, he wasn't sure if it would be possible with another topic such as Electricity with its heavy content load. He felt he could not afford the luxury of teaching for understanding when there were 'so many concepts to get through' in preparation for the external examination.

It is OK with [Sight and] Light because all the ideas [can be taught through practical examples]. But when I get to Electricity, later in the year, I'll probably revert to my old ways because there are so many concepts to get through that are necessary for [the external examination].

A further influence on David's teaching was the test results at the conclusion of the topic, Sight and Light. Students' enthusiasm for the new approach led David to anticipate high success in the test. He was disappointed to discover that the results did not match this interest and involvement: 'I was quite upset when I saw their test papers: I wondered if they'd learned anything at all'. He admitted that he had difficulty framing questions to assess students' understanding and skills rather than testing their content knowledge. But he explained:

What it has really highlighted for me is the process of trying to get them to talk about what they have done and to write about it in their own words goes right back to primary school. Somehow they don't think they have to write about science; they think it's a matter of lots of multiple choice questions where they just fill in the gaps. The other way is to learn whole chunks of the text book. They think it's a matter of a right or a wrong answer and learning a few facts off by heart.

I'm starting to realise we really can't hope to change much just by changing the syllabus at grade 11. What we need to do is to change attitudes and [the] whole approach to how [we] do things in science. We have to start way back getting them to verbalise and write down their ideas. It's got to start in the earlier years.

David's hopes and dreams for teaching physics in the new way were constantly moderated by tradition. He was still required to meet certain content objectives and to prepare students for the external examination. Equally, David was limited by his pedagogical content knowledge. His efforts to change were consistent with and necessarily constrained by his pre-existing knowledge and skills, and patterns of practice. His knowledge about teaching the new syllabus proceeded gradually and hesitantly rather than by a sudden grasp of new pedagogies.

Experimentation

David believed that certain strategies were conducive to developing students' qualitative understanding of physics concepts. His experience told him that understanding was enhanced through talking and writing about experiences. He insisted that students describe and explain 'in [their] own words' when answering his questions in class and writing about their experiments and class activities. He helped students to develop ways of expressing their own interpretations of what was happening rather than relying on textbooks as the source of explanations. Indeed, at the beginning of the topic, David suggested that students leave their textbooks at home. He argued that 'too many students learn whole slabs of text and regurgitate them in tests and exams'. At times he pressed students to clarify meanings of words they were using. Such was his concern for students' ability to write about their experiences that he spent one lesson working with students on their descriptions of what happened to the light rays as they passed through the curved surface of the glass. He analysed students' writing, helping them to find clearer ways of expressing their ideas. At the end of the lesson, David resolved to invite one of the English teachers to the class to work with him toward better ways of communicating physics concepts. By placing this emphasis on students' writing and expression of ideas, David also showed that valuing students' understanding was an important part of the new approach to physics.

These new teaching strategies took more time than David expected. Faced with a trade off between content coverage and teaching for understanding, David decided to create more opportunities to experiment with the new syllabus. He decided that if he were to make progress he needed to free himself from the demands of the old syllabus.

It was really a matter of throwing the [old] syllabus out for the section on Light. It was a matter of making the decision with grade 11 this year that it would be a time for trialing things without the constraints of what I had to get through in [one year].

David found time in the program for his experimentation. His strategy was to disregard time so that he could set his own pace with his class. He explained that 'what I have done is basically to throw time out the window'. However he discovered that teaching for understanding was much more time consuming than the traditional content-driven approach.

I spent a lot more time on it than I would normally. In two and a half weeks I only covered what I'd normally teach in two and a half lessons.

In creating opportunities for experimentation he capitalised on the fact that he was the only teacher of senior physics and also the head of the science department. As the sole physics teacher in the school, he was not bound by the constraining patterns of expectations of colleagues to meet deadlines and goals or even to plan together. Indeed, he used his freedom to allow him to respond to opportunities as they arose in the classroom.

If there had been another teacher it may have been a little more structured; we would have sat down and mapped it out together. But I quite liked the way things happened and I didn't feel constrained by a structured set-up with someone else . . . What I enjoyed about trialing it was that I could go back and forth as I was trying different things . . . I could zig zag a fair bit, depending on how things went.

David's experience illustrates the problems that teachers face with doing things differently. Teachers' work is constrained by many factors allowing for little time for inquiry. Within these constraints, David managed to create for himself freedom to experiment — to 'zig zag a fair bit, depending on how things went'.

Biography

To understand more about David's teaching, it is necessary to examine his biography and experience and the patterns of practice developed over many years of teaching science. David was one of the state's most highly respected physics teachers. At the time of this study he was president of the state science teachers' professional association, chair of the professional development committee at his school and a member of a similar committee in his region. He had recently been awarded an international travel fellowship for research into science teaching. Throughout his teaching career David had been called on as an advisor, a consultant and a reference person by his employer and by professional associations. He had participated in the debates and discussions that gave rise to the syllabus reform and taken a leadership role during its implementation.

Clearly, David's history as a leader in his subject and in his school meant that he could confidently exercise the kind of professional autonomy required to experiment with, and to question his approach to, teaching physics. His connection with the syllabus changes through his role as a 'link' teacher enhanced his philosophical understanding of the changes. These understandings were verbalised to his students and modelled in his practice. David's predisposition to consider and critique his practice also opened up the possibilities for improvement and change.

Nonetheless, David struggled with the changes asked of him. His experience with the strategies of the old syllabus which emphasised rote learning of facts, algorithms and definitions was extensive. Over the years these strategies had become routine and well rehearsed involving materials, assessment tasks, experiments and activities that he knew to be successful. David's experimentation in opening up the activities, encouraging student discussion and connecting with context were grafted onto the fundamental elements of physics teaching as he knew it. The content demands of the syllabus remained important, the algorithm of Snell's Law still needed to be memorised and students were still expected to work through the laboratory session in a practised manner. However, David's knowledge about teaching physics did proceed in the course of this study. His superior content knowledge, his repertoire of instructional strategies and his past history as an agent of change do much to explain the growth of his knowledge.

CONCLUSION

In David's teaching we see his past, present and future coming together. David's biography explains both his respect for tradition and his predisposition to change his practice. Steeped in the tradition of content coverage and the application of mathematical formula, and overshadowed by the spectre of the external examination, it is not surprising that David struggled with some of the new strategies. However, his commitment was also enhanced by his earlier involvement in syllabus development and he had a stake in making the program work. David's experimentation with his practice, though constrained by his past, contributes to defining a new future. He managed to expand his repertoire of instructional skills and strategies by tinkering (Huberman, 1992) with his practice within the confines of his classroom. In this way his knowledge about the new program, and hence his capacity to teach differently in the future, developed by the gradual extension of his horizons of understanding (Louden, 1991).

Teachers' learning, we argue, takes place in the territory between tradition and change, as the gap between current horizons of understanding and new experiences is bridged.[5] David's case, for example, represents significant learning as he moved, albeit gradually and hesitantly, between the old and the new. Perhaps we should not be surprised, for this is an account of a teacher in command of the craft and content of teaching science. David is an experienced teacher whose professional expertise and teaching patterns have been refined through many years of teaching science to diverse groups of children. He is also a senior person with strong associations with the new syllabus and the authority to experiment with new teaching strategies without fear of contradiction. In many respects, the combination of David's knowledge, skill and confidence within this particular curriculum, classroom and school setting appears conducive to teacher learning.

Not all teachers respond to change in the same way. Sometimes, the gap between old and new is so great that teachers fail to make the transition. What happens to teachers' learning in such cases? For example, what might have happened if David had been less confident in his content knowledge? What if he had been in a school where 'I'm just experimenting' was less acceptable or in a large physics department where he was required to teach to a common test? What would be the consequences of low test scores in a school with high academic expectations? These questions highlight the enormous range of conditions under which teachers' learning is expected to take place. In the following chapter, we begin to explore some of these conditions by describing the work of Mr Ward, another physics teacher who responded differently to the same syllabus change.

NOTES

1. Connelly and Clandinin (1988) use the term 'directionality' to describe how history shapes situations and how situations point to the future.

2. This chapter was jointly authored by Helen Wildy and John Wallace. Some parts of the chapter are reproduced, with permission, from a 1994 article by the same authors which appeared in *Research in Science and Technological Education, 12*(1), 63–75 (Abingdon, UK: Carfax).

3. Jackson (1995) argues that the object of fieldwork is not to represent the world of others, but it is a mode of using our experience in other worlds to reflect critically on our new experience. This view is consistent with the Dewian (Dewey, 1958) idea that knowledge is a process of coming to know (*erkennen*), rather than a body of received findings.

4. These theoretical ideas have received substantial treatment in other places. For example, on change, see Waller (1932), Fullan (1991, 1993), Hargreaves (1994), on tradition, see Blythe (1965), Louden (1991), on experimentation (and inquiry), see Huberman (1992), Cochran-Smith & Lytle (1993) and, on biography, see Butt and Raymond (1987), Connelly and Clandinin (1988), Goodson (1992), Witherall and Noddings (1991), and Schulz (1997).

5. For an extended discussion on how teachers' knowledge arises out of action, see the review by Fenstermacher (1994).

4. SUBJECT MATTER AND MILIEU: MR WARD

One of the problems with building theory based on close-up narrative accounts of teachers' work, such as the accounts of Johanna's and David's teaching in Chapters 2 and 3, is that they create a selective vision of the events they set out to study. There are several pressures which combine to create this selective vision. Perhaps researchers who plan to spend extended periods observing particular teachers tend to choose teachers with whom they feel comfortable. Hargreaves (1996) has argued that this has led researchers in this field to choose teachers who are humanistic, rather than conservative or politically radical. Whatever the researcher's perspective, however, there may also be a tendency for narrative case studies to reflect the common ground between the researcher and the teacher. The experience of close involvement with collaborating teachers leads researchers to 'walk a mile in their shoes', and may lead researchers to come to see the world somewhat as their collaborating teachers do.[1] Moreover, the ethical requirement to negotiate the final text of a study with the collaborating teacher may have the effect of emphasising events and interpretations that are viewed by the participants as positive rather than problematic.[2] Finally, the pressure on researchers to conduct field work in sites where students, teachers, and administrators are willing to welcome the observations of outsiders may tend to over-represent sites where participants are more than usually compliant.

With all these pressures towards a comfortable and selective vision of teachers' work, researchers working with narrative case studies must be rigorous in their pursuit of disconfirming evidence. It is not sufficient to build our understanding of teachers struggling towards constructivist science teaching on a series of safe cases such as the one in Chapter 3. What, for example, if David had been less welcoming to us as researchers, or less embracing of the syllabus changes? What if he had been more directive in his approach to teaching or held views which were counter to our own? What if he had worked in a school which would not tolerate the false starts and hesitation that characterise thoughtful experimentation by teachers? How useful are the categories of change, experimentation, tradition and biography in describing the work of other science teachers? What other issues need to be considered? In this chapter, we describe the work of a teacher with whom we expected to have less in common.

As we had done with David, we[3] approached Mr Ward with the aim of finding out how an experienced teacher would adapt his teaching to incorporate the teaching and assessment strategies promoted in a new physics syllabus based on constructivist principles. We visited Mr Ward in his classroom over several weeks, observing his lessons, talking with his students about their experiences and having lengthy conversations with him before and after each lesson. We also conducted formal interviews with Mr Ward and with members of his class. When we entered Mr Ward's classroom, we expected that the new syllabus, incorporating new assumptions about physics knowledge and how it is acquired, would challenge Mr Ward's beliefs about what he knew and could do well. We understood that the process of becoming a constructivist science teacher would involve the teacher in reconstructing his own knowledge of science and science teaching.[4] However, we were not entirely prepared for what we found when we went into Mr Ward's classroom.

MR WARD'S PHYSICS CLASS

Like David, Mr Ward was teaching the new grade 11 physics syllabus based on constructivist principles. Before we entered the tiered lecture theatre on our first visit to Mr Ward's class, he explained to us that physics was taught in five 45-minute lessons during a six-day timetable cycle. Because of the shortness of the lessons and the need to 'keep in step' with other teachers, not only for lessons but also in the timing and content of tests, he felt that he 'couldn't play around' with his teaching. The domination of the time frame became evident from the start of the first lesson we observed. Mr Ward introduced the new topic for the semester by distributing a program headed:

Unit: Movement And Electricity
Area Of Study: Movement
Context: On Your Own Two Feet

He explained the program in terms of the time allocated for each section and the sets of exercises to be completed, expressing the concern that there would be 'not a lot of flexibility'. Next he distributed a Student Outcomes sheet to each student, stating: 'This is what you have to be able to achieve, in other words what you'll be tested on'. He pointed out differences between this course and the 'old course' in terms of organisation and then gave some indications of where they might save time. For example, he told his students:

We'll break this topic into two sections for assessment. In the old course this was five units but we've cut out a lot. There's not so much on vectors in the new course. You can see that 'Position' has only four lessons and I hope to do it in less because you cover this in chemistry. We can save ourselves some time.

Mr Ward passed around a third sheet headed International System of Units (SI) and, by way of introduction, told the class 'I assume you know all this'. As he explained the tables of Unit, Symbol, and Quantity and Prefix, Symbol and Meaning, he reassured the students that 'the tables booklets you're given in the [university entrance examination] contain these prefixes so you don't have to learn them'. In the first 10 minutes of the first lesson on the topic he had made explicit the time frame and clarified the content required for successful completion of the topic. He had also begun to make clear what might and might not be expected of this topic in the external examination at the end of grade 12. He also talked about 'exam etiquette'. On the board he worked a numerical example based on the International System of Units. When one of the students, Fiona, asked whether it was better to write the answer as $(1 \times 10^2)^3 \, m^3$ or as $10^6 \, m^3$, Mr Ward explained the need for 'creating the correct impression':

The question is: What is a polite impression to create? What do you write to impress the examiner? How do you use the terminology to let the examiner know what sort of student you are? He will want to know: Is this an A student? It's about creating the correct impression. No, it's definitely not polite to leave your answer as $(1 \times 10^2)^3 \, m^3$ rather than $10^6 \, m^3$.

Posing each question and answering it as he went, Mr Ward proceeded through the set of numerical exercises. When he came to: 'Express 0.002A without the use of prefixes' he cautioned the class to 'watch out for dirty tricks from the examiner' and explained that 'of course, this one doesn't have to be changed at all'.

When the sheet of exercises was finished several students noted 'We've finished a sheet — in one lesson!' Mr Ward laughed with them as he distributed another sheet of exercises; this one was titled Use of Scientific Notation. Again, Mr Ward worked through the examples with the class. He helped them 'lock' their calculators into scientific notation by identifying the various buttons to be pushed on the various models of calculators used in the class. By then it was the end of the 45 minute lesson and, as the students left the room, Mr Ward explained:

The students don't necessarily want to study physics to do physics but to get into university. . . . They're interested not as much in physics as in their [examination] aggregate. So everything that happens in class must bear directly on that. Anything else is seen as a wasteful digression. We are all here to get good [examination] results.

We observed other instances when Mr Ward focused on examination preparation. Sometimes this took the form of 'playing it safe' in relation to changes to the syllabus. Mr Ward began the second lesson of the topic by referring to a sheet headed Uncertainty in Measurement saying:

I want to revisit the issue of uncertainty in a more sophisticated way than we do in our chemistry classes. We need to be more precise in physics. [The examiners] expect you to be able to work out errors.

He proceeded to perform calculations to illustrate the difference between 2.0000g and 2g. He told his class that 'in the old course we used to spend a lot of time doing detailed analyses of errors'. By way of justifying having spent class time doing just that, he explained: 'We don't have to do this any more but we'll believe it when we see the [new] exam papers they set next year'. So in the meantime he was being cautious, shielding the students from the possibility that he may have been misled. He did this by showing students how to calculate percentage error, and the difference between absolute and relative error, even though such a quantitative approach to the concept was not in the new syllabus. He continued to explain the conventions in the use of significant figures. Students were clearly interested, asking many questions and generating their own examples for Mr Ward to work on the board. After some time he stopped the calculations, saying 'If I had more time — which I haven't — I'd show you how all this really worked'. And he returned to a worksheet, talking through the questions with the students calling out answers. In returning to the worksheet Mr Ward drew attention to what is worth spending time on, and what is not: only if it is explicitly 'in the course' can he justify taking class time.

Another way Mr Ward helped his students prepare for the final examination was by distinguishing the level of difficulty of questions in set exercises. Several lessons later as he distributed the worked solutions to a set of problems, he explained:

Some questions in here are of the extension category — too hard without help, like numbers 5 and 11. We used to teach the techniques for solving number 11. You'd be very unlucky to strike one of these in grade 11 or 12 even. Question 6 is also hard. You should be able to work through it but don't think you're thick if you can't do it.

Again, he made explicit the boundaries of the new syllabus as he referred to data collected in the previous week's practical activity:

We spent time on Friday doing mathematical manipulations of data we collected on Thursday. Now you could be asked to do this. But you are expected to be able to interpret the graphs rather than just manipulate the data. So we need to look at the velocity-time graphs and the acceleration-time graphs.

During the previous week the class engaged in a practical activity designed to generate velocity and acceleration from distance-time data. The lesson took place in the playing field; students sprinted on a 100m track and were timed at 10m intervals. Despite having the data collected from students' own sprints, Mr Ward returned to the sheet of Typical Graphs of Uniform Motion and explained that 'in this course we are really only concerned with situations of uniform velocity and acceleration'. When Rebecca said: 'I don't understand these graphs — I mean, how to use them or just what they mean', Mr Ward reassured her:

You need to be able to recognise them. You can commit them to memory unless you're one of those mathematical types who immediately converts a set of conditions into a graphical form. Don't worry too much for now though. We'll spend the next two lessons applying them.

Using a complex velocity-time graph drawn spontaneously on the board he explained why the slope of the line on a velocity-time graph gives acceleration and the area under the curve gives displacement. He did this first using general terms, v_1, t_1, t_2, t_3, etc. Although students were listening attentively, it was not clear to Mr Ward how much they understood. However, when Mary asked 'why did you put the line in there?' it became apparent that a more concrete example was needed. Mr Ward took out a tennis ball, his 'standard prop for this topic'. With illustrations of throwing the ball in the air, Mr Ward explained positive and negative velocity, going forwards and backwards and positive and negative acceleration. Then he returned to his graph on the board, replacing the general terms with numbers and continued to calculate displacement and acceleration for the various sections of the graph.

The following segment of conversation illustrates how Mr Ward continued to define what was worth knowing about in terms of what was in the syllabus.

Sarah: I hate to be a pain but I still don't understand what the negative slope
            means.
(Mr Ward explained again, using his diagram on the board.)
Sarah: Now I get it.
(Mr Ward continued with his calculations of velocity and acceleration.)
Sarah: Hang on. What if that isn't a triangle? What would you do then?
Mr Ward: If it's uniform acceleration, it will be [a triangle]. If not, you have to use
            calculus or count squares [to find the acceleration]. But that's not in
            your course so you wouldn't be asked.

We watched the teaching strategies Mr Ward used in his classroom. His most common mode of teaching was to talk to the class. He did not ask questions. However, he willingly and patiently answered all questions from students, with equal respect, as though even the most elusive concepts would eventually be understood by all students given sufficient time and explanations. And the students frequently asked questions. Although there was no planned interaction between students, they did talk to each other as they carried out tasks while Mr Ward was talking. That their teacher 'doesn't mind questions' and creates a 'friendly atmosphere' contributed to students' enjoyment of physics lessons. They frequently commented that Mr Ward's relaxed style encouraged them to ask questions 'even questions that don't have much to do with it, kind of "on the side" questions'.

Apart from practical activities, there was only one lesson style that deviated from this: it was what Mr Ward called the workshop. He described this type of lesson to us as:

... giving the students plenty of opportunities to come to grips with the ideas. I need to give them time to play around with the ideas through the problems. There are exercises from [the text] that are good and I have lots of sheets of my own that are useful, too.

This is how he explained it to his class:

I'm tired of doing all the work while you're sitting listening or chattering. I've got a couple of sheets for you to do. You can ask your neighbour for help or me. I'm collecting them at the end of the period. It's a way for me to find out how well you're going. We've got a test coming up in a couple of weeks time so I need to know if you are getting ready for it.

Immediately the students started to work, mostly on their own, through the sheet of problems headed Rectilinear Motion Test, checking answers with each other as they went. When Jane asked Mr Ward for help, he stood by her and explained how to do the problem, giving the solution without discussion or engagement, and then continued talking in his normal level of voice. Another student, Susan, complained: 'Don't tell us, Mr Ward. If you tell us all the time, how can we learn?'

UNDERSTANDING MR WARD'S TEACHING

This was the question we also asked. What is the connection between Mr Ward's practice and the knowledge he has and uses? How does this connection lead to a different kind of classroom from that of David? How do the theoretical notions of change, tradition, experimentation and biography apply in Mr Ward's case? What other dimensions help our understanding of the differences between Mr Ward's and David's learning?

Change

In Chapter 3, David's case provides evidence that teacher knowledge develops gradually and hesitantly, rather than through sudden leaps of insight. David struggled with the new approach to teaching physics as he reached the limits of his pedagogical content knowledge. He proceeded to a new level of competence with his physics teaching by creating freedom to experiment with the new syllabus. Like David, Mr Ward approached the new syllabus with reservation but a willingness to try some of the suggested strategies. He explained how he had experimented initially with a more context based approach.

At the beginning of the year I did things differently but I didn't feel comfortable with them . . . When I tried to be totally context based the structure of the subject disappeared.

The consequence was that students started to develop 'very negative attitudes towards physics' because 'they expected to get 80s and 90s for tests and when they got 30s they were freaking out'. Students began to change subjects in greater numbers than Mr Ward had previously encountered.

Well, we certainly lost students who said: 'That's enough for me; I'm going'. That doesn't usually happen. In the past you could count on the fingers of one hand the number of students you lost. In the previous 10 years you'd lose one or two a year at the most. And here we were losing three or four in the first term.

Although the new syllabus was intended to make physics attractive to a wider range of students, Mr Ward believed 'we've achieved precisely the opposite effect'. In his view the mathematical demands of the old syllabus were less but the intellectual demands of higher order skills like synthesis are 'much more demanding and much more stressful on the student than the old demands'. Mr Ward's fear was that once students realised the level of difficulty of the new syllabus they would 'leave physics in droves'.

Faced with the prospect of losing more students, Mr Ward opted for a more conservative approach, recognising the importance of context in the new syllabus but building on teaching techniques which had proven themselves over the years.

I felt uncomfortable not knowing where we were going so in the end I moved back to a more structured approach but emphasising the context. I found it much better simply to emphasise the context more but still retain the structure.

Mr Ward's approach was to 'extend what [he] was already doing' rather than make significant changes to his teaching strategies. Rather than begin with the physics context, the approach proposed in the new syllabus, Mr Ward chose to incorporate the context into the formal structure of physics as he knew it. This approach allowed for subtle changes to his practice while retaining his integrity as a physics teacher and the confidence of the students in his class. While Mr Ward's case differs in several respects from David's, for both teachers change proceeded in similar fashion — by way of hesitating steps forward, at times retreating back to familiar and tested ways of teaching before venturing forth once more.

Tradition

For Mr Ward, as for David, tradition stood as a powerful moderator for the changes he made in his practice. Mr Ward relied on the authority of his 20 years experience[5] to make decisions about what was important and what was not. When some students withdrew from physics, it was enough to convince him that teaching Light through the context of photography was inappropriate, time consuming and ineffective. He 'didn't feel comfortable' with constructivist teaching strategies because they took time away from doing 'the work as set down' in the syllabus. Mr Ward understood that his main goal was to prepare his students for entry to university study. He recognised that his responsibility to his students was to introduce them to the discipline of physics and 'to get them through'.

Looking at the class that I've got there I would be surprised if there were any who weren't at university in 18 months time. Lots of them won't do physics but most of them will be at university. Most students who do physics are doing it to keep their options open and the options involve courses requiring knowledge of formal physics and those requiring [examination] scores of a certain level.

As a teacher with a history of success in preparing students for examinations, he was confident that his traditional approach to teaching physics would produce results:

There is far too much in the course for the students to reach the level of competence they feel confident with in the time available. . . . to do things well rather than merely adequately . . . to work over areas to get confident, so they feel confident and I feel confident that they've really grasped the idea.

Tradition was also important in Mr Ward's school. Examination success was celebrated and promoted. Parents chose to send their children to the school because of the high rates of university entrance. As a part of this culture, Mr Ward's practice was bound by the traditional values held by the wider school community. These values reinforced particular kinds of teaching practices — coaching to the examination, algorithmic practice and close attention to the syllabus. Mr Ward was a successful teacher in this environment, highly skilled and highly respected. So the traditions that served Mr Ward so well in the past, bringing him recognition and respect, also served to moderate his ongoing practice. In this respect, Mr Ward and David have much in common. For each of these experienced teachers, tradition enables them to teach physics with authority and certainty, and defines their horizons of understanding.

Experimentation

In Chapter 3, we described how David managed to create space in his program to experiment with the new strategies. Mr Ward's circumstances were more constraining than David's. His hesitating efforts to incorporate a more contextual approach by teaching Light through the context of photography did not meet with immediate success. Mr Ward was sceptical about the approach and perceived a certain level of student resistance. Like David, Mr Ward found that teaching through contexts was more time consuming than the traditional approach and that there was no guarantee that the students would arrive at the formal principles of physics. While David was willing to persevere with the contexts approach, Mr Ward felt that the risks involved were much greater than any potential benefits.

A more successful example of Mr Ward's experimentation was what he called the 'comprehension exercise'. Mr Ward insisted that his students needed to be confident not only in the ideas he was developing but also in the methods by which they would be assessed. The 'comprehension exercise' was one of his strategies to ensure that his students were well prepared for the examination. He explained how he did this:

You've got to have assessments of comprehension . . . For each topic we give them practice at those things that are now expected of them. We give them a test which is a comprehension exercise so they're given a chunk of text which is on the theme they're working on and they are asked questions to see their understanding of the text and to apply their knowledge from the area they are working on to the text to make meaningful physics out of it. That takes some practice. They've had exercises which we've done in class and they've had homework on it. We've actually had an assessment item like that on each section as we've done it. Because that kind of thing is going to be really part of their exam structure now they have to become reasonably competent at doing it. If you don't train kids to do things they never learn to do them.

What Mr Ward did with the 'comprehension exercise' was to build on his usual practice of helping students with assessment structures to incorporate a new kind of assessment exercise. This example of successful experimentation did not involve a large degree of risk taking but rather a small amount of tinkering with well tested methods of teaching.

Biography

Mr Ward, like David, was an experienced science teacher. And, like David, much of Mr Ward's practice can be explained in terms of his biography. He was head of chemistry at his school and had taught physics at high school level for more than 20 years. Previously he had been a university tutor and lecturer for several years. He was well known and highly respected for his teaching, both in his own school and within the wider community of science teachers. His physics content knowledge was extensive in breadth and depth. He was confident in a set of teaching strategies that worked for him. Confronted with a major syllabus change, he experimented with a number of the proposed teaching strategies. Some of these strategies, he rejected as inappropriate because they did not produce the evidence of learning he had anticipated. He did not feel comfortable and neither did some of his students: they failed to learn the basic physics concepts, lost confidence and left the course. To Mr Ward this was enough to convince him to revert to his former well practised strategies.

Mr Ward's biography helps us understand his decision to 'move back to a more structured approach'. His routines had worked well in the past. He made clear to students the boundaries of the content by identifying what was, and was not, in the course. By attending closely to the time schedule he ensured all content was covered, even though the tight time frame did prevent him from 'playing around with his teaching'. He instructed them in the conventions and protocols of physics as a discipline. He spent considerable time on examination etiquette and in developing students' confidence in new assessment strategies such as comprehension exercises and oral presentations. Mr Ward recognised that at times it was necessary for students to take short cuts, for example, to commit graphs to memory. In making explicit these rituals and routines, learned over many years, he also acknowledged that, for his students, the importance of physics knowledge lay largely in its role in generating high examination scores for university entrance.

Mr Ward's biography includes his experience in dealing with similar classroom issues and problems of practice in many physics classrooms. However, we have also argued that his work, and his world view, is also shaped by his adherence to certain traditions about the teaching of physics in certain contexts. Two traditions in particular — the structure of the subject matter[6] and the milieu of the classroom and the school — are foregrounded in this case.

Subject Matter

Mr Ward wanted to convey to his students the understanding of, and feeling for, physics as a discipline. He was concerned that students 'learn to live within the discipline' with its structure of recognised and established protocols and conventions, without which he felt the structure of the discipline 'disappears'. Further, he did not believe that students, alone, could find and shape the structures: it was his responsibility to 'construct the framework' of the discipline for his students. Without the conventions of the discipline, too, 'the context approach degenerated quickly into a version of discovery learning where it took an awful long time to discover an awfully small amount'. When Mr Ward had used photography as the context to study Sight and Light, this had caused problems.

You soon got lost in the complexity of photography. The principles of Light came three or four days after we started so we had spent all that time talking about things some of the students understood because they had some background in photography and others didn't. The end result was none of them were particularly good at the basic ideas of Light.

Mr Ward rejected a totally context-based approach when he found the structure of the subject, with its recognised protocols and conventions, disappeared. Mr Ward is not alone in his view. White (1989) argues that the scientific discipline is a social mechanism that gives order and structure to the study of nature, itself 'an inherently untidy experience' (p. 191). Without its special use of language to interpret the world, science cannot be understood (Martin, 1990). The language of science is used to classify, decompose and explain; its protocols and conventions are used to define and structure the discipline. This technical language involves generic structures, like reports, explanations, definitions and experiments, to construct the content of science and a scientific world view. To downplay scientific discourse is to downplay the science that is taught because 'science is unthinkable without the technical language it has developed to construct its world view' (Martin, 1990, p. 115). Others such as Costa (1993) argue that school science with its specialised language and agreed procedures, is a rite of passage inducting students into membership of a scientific community. Elsewhere, we have argued that good science teachers place high regard for the conventions and structures of the discipline and that teaching for understanding is defined in terms of helping students to accept and use the language and protocols of science (Wildy & Wallace, 1995).

Milieu

Another tradition shaping Mr Ward's teaching was the importance of the school milieu. It was clear, for example, that the students in this school were comfortable with his mode of teaching: lots of teacher talk, clear delineation of course content, practice in examination techniques, and the absence of 'wasteful digressions'. When Mr Ward attempted a more inquiry context-based approach, students failed to learn the basic concepts. It is likely that these strategies, designed to promote understanding, had limited impact because the students held competing beliefs about teaching and learning.

When teachers attempt to apply constructivist strategies designed for personal and social meaning making, students frequently adopt passive roles for themselves in line with their transmissive views of teaching and learning (Gunstone, 1990, 1992). Students may simply deny the legitimacy of the strategies or, at best, undertake tasks in ways that minimise the demands made of them. And there seem good reasons why they do. Why should students make knowledge theirs when they are confronted with ready-made results to pre-constituted knowledge? Why should they try to engage in conceptualising when most of the time they have been passively involved in finding solutions that are already known? Students opt for ready-made solutions rather than making their own meaning precisely because they know a lot about the school system and the type of science knowledge it promotes.

Mr Ward knew a great deal about the school system and the type of knowledge it promotes. He shared with his students an understanding that school physics is situated within this milieu, not separate from it. He built on his students' prior knowledge of the milieu and helped them achieve success within it. Mr Ward understood and accepted his students' goals and reassured them he would help them do sufficiently well in the subject to guarantee examination success. Maintaining trust seems to be a critical ingredient in the unwritten but powerful agreement the teacher makes not only with his students but also with their parents and the school community.

The interplay of students' personal agendas with the goals and values of the science curriculum, the teacher and the school are powerful factors in Mr Ward's physics class. Equally important are classroom contextual factors, particularly the task, authority and evaluation structures, which influenced the way students responded to Mr Ward's strategies. The students understood that the evaluation procedures emphasised competition and external reward rather than deep understanding. They knew about the importance of working within an allocated time frame, following instructions, and offering clear and definite answers. For the students, getting it done was more important than thinking it through. Mr Ward understood the importance of these social forces and adjusted his teaching accordingly.[7]

For their part, students accepted teacher talk, examination coaching, algorithms, and the absence of real world experience in their physics lessons. This was the image of teaching with which they were familiar and comfortable. In the eyes of the students, their parents and the school community, Mr Ward was a good teacher and it was evident from his relationship with his students that he had their confidence. That confidence was tested when he experimented with his teaching at the start of the year. He attempted to use strategies that did not fit with his students' beliefs about teaching and learning. By talking about the routines and rituals he used, Mr Ward steadily rebuilt his credibility and by showing his students that he knew what they expected, Mr Ward regained their trust that he would help them achieve their goals.

CONCLUSION

Earlier, we admitted to some surprise at what we saw in Mr Ward's class. Whole class instruction, algorithmic practice, teaching to the examination, content coverage — these pedagogies appeared to be quite foreign to the spirit of the new syllabus and unlike what we had observed in David's class. Yet, as we came to understand what was happening in Mr Ward's class, we recognised many of the forces we had previously identified in David's practice. Change, experimentation, biography and tradition operated for Mr Ward, as they did for David, to shape his approach to the new syllabus and define the limits to his horizons of understanding. Nonetheless, Mr Ward's teaching was different from David's in important ways. Mr Ward's long experience studying and teaching physics at high school and university level led him to a particular view of the importance of the formal structure of the discipline of physics. This view led him to believe that physics could not be effectively taught through a contextual approach. It is also clear that Mr Ward's teaching was centrally influenced by other realities of schooling — students', parents' and probably colleagues'. Perhaps a major reason for the difference between Mr Ward's and David's practice is that these realities are more explicitly stated and understood in Mr Ward's school than in David's school.

What Mr Ward adds to our understanding of science classrooms is an appreciation of the importance of the content of science and the cultural context of schooling in shaping teachers' horizons of understanding. Mr Ward's case provides an example of the teacher having a clear and consistent view of the subject matter and teaching goals congruent with the needs and aspirations of the school community. Elsewhere, we have argued that these are essential characteristics of good teaching (Wildy & Wallace, 1995). But is Mr Ward correct in his interpretation of the nature of physics? Have we been overly sympathetic to a teacher who adheres to his conservative ways in the face of innovative curriculum change? What about teachers who are less clear or less confident about the content than Mr Ward? What happens when the 'other realities' of the classroom are more diverse, less explicit, or even at odds with those of the teacher? Whose reality counts in these situations? We finish this chapter as we began, by advocating that narrative studies of teaching should not be confined to safe accounts of pedagogically strong and/or innovative teachers. In Chapter 5, we add our observations of another teacher, Ms Horton, to those of Johanna, David and Mr Ward, to build our theoretical understandings of the nature of science classroom.

NOTES

1. Hatch and Wisniewski (1995) cite Petra Munro who warns against what she calls the 'romanticization of the individual' in narrative research. Goodson (1995) argues that researchers have an obligation to situate personal stories within socio-cultural and political contexts.

2. Wolfe (1973) refers to this problem as part of the dilemma researchers (including journalists and case reporters) face in trying to maintain integrity in their research while, at the same time, safeguarding the well-being of participants.

3. This chapter was jointly authored by Helen Wildy and John Wallace. Some parts of the chapter are reproduced, with permission, from a 1995 article by the same authors which appeared in *Journal of Research in Science Teaching, 32*(2), 143–156 (New York: John Wiley & Sons).

4. For a discussion of the constructivist paradox, see Louden and Wallace (1994).

5. Munby and Russell (1993) used the term 'authority of experience'.

6. For elaborated treatments of the importance of the subject matter tradition on teachers' lives, see Goodson and Marsh (1996), Grossman and Stodolsky (1994), and Siskin (1994).

7. Costa (1997) uses the example of honours chemistry to show how the teaching and learning of high status science subjects is shaped by the sociocultural precepts of the school and wider society. A further example, of university-entrance physics, is provided by Wildy, Louden and Wallace (1998).

5. MULTIPLE REALITIES: MS HORTON

This chapter argues there are many realities of science classrooms, as events are experienced and perceived differently by teachers and their leaders, and by teachers and students. Narrative accounts which focus primarily on the teacher's voice sometimes fail to recognise that these realities are often divergent and dissonant in nature (Hargreaves, 1996). Indeed, this has been a feature of our own writing in previous chapters where we have depicted the classroom through the teacher's eyes. Even in the case of Mr Ward, where we first addressed the notion of other realities, we proposed that he and his students had a singular view of the goals of classroom and of good science teaching. However, classrooms are more complex than that. We know, for example, that issues of culture and class affect science instruction and social mobility (Wildy, Louden, & Wallace, 1998). Hence, this chapter attempts to deconstruct the notion of a single classroom reality — often that of the teacher or the researcher — imposed onto other voices. In doing so, we attempt to adopt a 'contrapuntal perspective' (Said, 1993). Instead of adopting one standpoint and interpreting from that exclusive perspective, we are aware of the innumerable alternatives that might equally be adopted.[1]

We must be able to think through and interpret together experiences that are discrepant, each with its own particular agenda and pace of development, its own internal formations, its internal coherence and systems of external relationships, all of them co-existing and interacting with others. (Said, 1993, p. 36)

The narrative which follows is an account of characters and events in a grade 10 chemistry classroom.[2] The main character is Ms Horton, an experienced biology major teaching the topic of *Chemical Change* for the first time. Karl and Punipa are two students in Ms Horton's class — a non-streamed group from a working class neighbourhood. Karl has been achieving marginal C grades in science and Punipa is a borderline A student. Karl's parents are of Maori origin and Punipa's parents are from Cambodia. Mr Greg, a long standing head of science, was newly appointed to the school at the beginning of the year. Ms Horton's classroom was the site of a two-month study of science teaching involving several researchers including John, one of the current authors. During this time, the class was observed on four occasions each week, and several interviews conducted with the teacher, selected students and the head of department. Data consist of field notes and analyses, transcripts and dialogue journals with the teacher and the students. In this chapter, the data are used to construct several vignettes featuring Ms Horton, Karl and Punipa, and Mr Greg, the head of department. Following the vignettes, we offer our interpretation of the perspectives of the four main characters and raise several questions about science teaching and learning, and about research into science classrooms.

TEACHING CHEMICAL CHANGE

As a researcher, I came to this study of a grade 10 science classroom in an Australian high school as a 15 year veteran teacher and head of department in the same school system. As a former teacher I have a sound understanding of the craft and content of teaching high school science. I well remember the rhythms of the school year, the scope and sequence of the science topics, and the strategies I used to occupy the students and ensure that the content was covered in the time available. As a former head of science, I recall my interest in maintaining a cohesive science department, well organised resources and assessment structures, and clarity of purpose. I also remember that some topics were more important than others in the task of selecting and sorting students for further science study. One such topic was — and still is — called Chemical Change. This ten-week topic covers concepts such as ionisation, valency, ionic bonding, acids and bases, chemical formula, metallic structure and molarity. Many science teachers consider this unit marks the crucial moment in the life of a student when high school science gets serious. Teachers usually teach this topic with more intensity, more urgency, more focus, in the knowledge that students' grades will be used as a guide to select those who will study chemistry and physics in the senior years of high school.

The Pop Test

One of the early lessons for the topic of Chemical Change was on the properties of acids. After conducting an opening discussion about household acids and the importance of the scientific method, Ms Horton distributed instructions for an experiment involving the reaction of acids with metals and the collection and testing for hydrogen gas (the pop test). The students were asked to work in groups or pairs. During the lesson, Ms Horton moved around the classroom, helping students with the experiment and asking questions about what they were doing. There was a fair amount of movement and noise in the room and Ms Horton was concerned about matters of safety and control. Several times, she interrupted the class to explain the worksheet, ask students to stay on task and point out the dangers in the handling of acids. Some students had difficulty in collecting sufficient hydrogen for the pop test because the acids provided were at low concentrations. This caused the students some frustration as many tried several times to collect enough hydrogen to get it to pop. Ms Horton was occupied helping a few of the students (mostly those at the front of the room) while others struggled with the experiment. With around five minutes remaining, Ms Horton asked the students to pack up their equipment. There was no time to recap the outcomes of the lesson.

Karl, who sat at the back of the room, seemed only marginally engaged in the lesson. During the opening discussion, his response to one of Ms Horton's questions about which household products might contain citric acid elicited no reaction and so he began to doodle on his notebook and talk quietly to his neighbours. When the experiment started, Karl found that he had missed out on receiving a copy of the instructions so he spent 10 minutes wandering around the room asking other students for the instructions. When he did manage to borrow some instructions, he appeared to have difficulty understanding the procedure and spent the rest of the lesson mixing acids and metals haphazardly and showing the mixtures to his friends. Ms Horton came to help once or twice. Karl seemed interested in what she had to say, but she could only spend a limited time with Karl and when she moved on, he resumed his previous activities. He filled in the period in this way. Without attracting attention from Ms Horton he played with the chemicals, talked with his friends in the back row and occasionally visited students in other groups to see what they were doing.

Punipa sat in the second row of benches from the front of the room. She listened intently to Ms Horton as the experiment was being introduced. Like Karl, she had some difficulty understanding the instructions because the equipment provided was slightly different from that described on the worksheet. In addition the acids were labelled with common names while the worksheet used chemical names. She watched what other students in her vicinity were doing and carefully added acids to the metals, noting her observations. Some of the acids, like lemon juice, did not seem to react with the metal. She found that when she tried the pop test it did not work, so she waited for Ms Horton to come and demonstrate. Even Ms Horton had to repeat the gas collection phase and try the test two or three times to produce a pop. Punipa persisted with the experiment but often looked up to see what other students were doing. For much of the time she appeared to be waiting for guidance or trying to figure out the purpose of the experiment.

In the next lesson, Ms Horton focused on the problems with the experiment. She asked students to identify the reasons why the experiment did not work. Several students offered reasons such as poor technique, weak acids and problems with the worksheet. Ms Horton then asked the class to design another experiment to discover one of the properties of acids. She distributed a worksheet which the students were asked to use as a framework to design their experiments. The students worked in groups as Ms Horton moved around the class giving help. Karl, Punipa and many of the other students had difficulty with the language of the design task which introduced new terms for them such as phenomenon, factor, hypothesis and variable. As with the previous lesson, Ms Horton interrupted the class at various times to provide an explanation of these different terms. Some groups appeared to make progress on the task but others seemed confused about what they were supposed to do. At the end of the lesson, Ms Horton handed out a homework assignment on common acids and bases.

Watching this sequence of lessons and observing Ms Horton and her students, I was puzzled by several things. The first was the apparent lack of urgency in Ms Horton's teaching. She had taken two periods to teach one aspect of the properties of acids — that they contain hydrogen. Even then, I wasn't sure what the students had learned. I appreciated her concern that the students needed to understand what they were doing but wondered whether she would be able to cover the content of the topic if she continued at this pace. Ms Horton seemed overly concerned that the students had not handled the experiment well and persisted with her emphasis on the scientific method. There did not seem to be a clear connection between the first lesson — a structured practical — and the second — where the students were asked to design an experiment. The main teaching point of the lesson sequence — that acids contain hydrogen — had been missed somewhere along the way. I also wondered about Karl and Punipa. Karl seemed to be interested enough in the lesson but the absence of clear instructions and focus meant that he was easily discouraged. Punipa was attentive but she struggled to make the connections between properties of acids and what actually transpired during the lesson.

'Important Stuff'

For the next two weeks, the pace of Ms Horton's teaching proceeded in a similar manner. Usually she introduced a concept through a structured practical activity and in a follow up lesson the students finished the activity. Ms Horton then conducted a class discussion to consolidate the students' understanding. Invariably, there was homework related to the activity.

About two weeks after the pop test experiment, Ms Horton came to class with a series of overhead transparencies and an associated worksheet. This was the first time that I had seen her use the overhead projector. She commenced the lesson by saying: 'We are going to do a lot of important work today because your test is next week'. She introduced the concept of neutralisation and gave several examples of acids and bases and how they reacted to form a salt and water. She wrote a number of chemical equations on the blackboard, asking questions as she proceeded. She explained how to use valencies to generate the formula of an ionic compound, making extensive use of the blackboard to show how to balance these formula. As a final activity, she distributed the worksheet and asked the students to complete a matrix of ionic formula.

Setting the students to work, she reminded them that 'this is really important stuff. This is the building block — the Lego for a lot of other work'. As students worked she moved about the room giving assistance with more urgency than I had previously noticed. On several occasions she reinforced the message that this was important material. To one of the students who was having difficulties, she said: 'Hang in there otherwise you will close off your mind and you will miss it'.

Karl began the lesson by copying the definition of neutralisation from the overhead transparency. He didn't complete the definition before Ms Horton had replaced the overhead with another so he leaned across to his friend and tried to copy what he had written. The next overhead contained a list of bases and Karl tried to copy this list in his slow hand. When Ms Horton removed the overhead, Karl put up his hand, presumably to tell her that he hadn't finished copying the notes. When Ms Horton failed to respond, Karl put down his pen. As the teacher gave examples of neutralisation, he chatted quietly with his partner and then looked forward hunched over the desk. When the students laughed about something that Ms Horton had said, Karl also laughed. At one point Karl raised his hand to answer a question but quickly pulled it down when his partner looked at him with disapproval.

When the worksheet was handed out, Karl looked at it for a few minutes before getting up from his seat to ask a student in another row what to do. He returned to his desk and attempted the first few formula. When he got to the more complex problems he looked for Ms Horton to see if she was available to help. When he saw that she was occupied he put his pen down and chatted to his partner until the end of the lesson.

Punipa also started the lesson by copying the definition of neutralisation, but unlike Karl she managed to keep up with the pace of the teacher's overheads. She did not raise her hand during the lesson but Ms Horton did ask her a question about acids in car batteries which she answered in a soft voice. She attended carefully as Ms Horton explained the formula for the acid-base reaction and wrote down the examples from the blackboard. Although she was sitting next to another student, Punipa hardly interacted with her partner while Ms Horton was teaching. When the worksheet was handed out, again, Punipa worked independently referring to her textbook for information on valencies and to check whether she was on the right track. When she got to the difficult problems, she had a brief conversation with her partner about what to do next. Ms Horton walked by at one point and told the girls that they were doing a good job.

This version of Ms Horton's teaching seemed quite different from the one I had seen earlier in the topic. Previously time had been used quite flexibly. In this lesson, time had been a more precious commodity. For example, in the pop test sequence observed earlier, Ms Horton allocated extra time to explore and reinforce ideas about the properties of acids. In this lesson, there was no extra time. Indeed, several new and difficult concepts — neutralisation, chemical equations, valency and ionic compounds were introduced in the space of a single lesson. Her teaching in this lesson was more didactic, more urgent and more insistent than that which I had previously observed. Ms Horton had forced the pace with her use of pre-prepared overhead transparencies and her strong messages to the students that this was 'important stuff . . . the building block for a lot of other work'. This lesson was also the first time that she had referred to the upcoming topic test.

I was also interested in the different responses of Karl and Punipa to this kind of lesson. Karl was obviously overwhelmed by the density of the content and the rapid pace of the lesson. He made some attempts to engage but was soon distracted by his peers and found other ways of filling in the time. Punipa, on the other hand, was engaged by the content in a way that I had not previously observed. She seemed to respond to Ms Horton's sense of urgency and remained focused on the task throughout the lesson.

Mr Greg and the Common Test

Puzzled by these matters, I began to delve into the reasons for the sudden change of pace in Ms Horton's teaching. One explanation was that Ms Horton was simply shifting into revision mode, a tactic often employed by teachers around test time. However, this was not old material but new and Ms Horton had not previously mentioned the topic test. There were clearly other factors involved in Ms Horton's sudden decision to change her method of teaching. One clue to this puzzle lay in Ms Horton's relationship with Mr Greg, the new head of science at the school.

According to Ms Horton, Mr Greg had been concerned about her progress through the topic. The two other grade 10 chemistry classes at the school — including a class taught by Mr Greg — were much further ahead. Mr Greg was worried that Ms Horton — a first time teacher of Chemical Change — was not covering the content of the topic. Several conversations had occurred between Mr Greg and Ms Horton about the content and pace of teaching the Chemical Change topic. At first Ms Horton managed to resist Mr Greg's pressure to move faster through the topic, arguing that her students needed more time to digest the material. In her words:

I originally went to negotiate with Mr Greg, and I think that a lot of the learning progress that we are having at the moment is becoming slower because the students are not having any chance to consolidate things at home. I would well imagine that not many of them actually go home and read what they've done or think about what they've done unless you ask them, in terms of science.

Ms Horton asked Mr Greg if she could delay some of the material until the following term. She asked him if she could look at some of the objectives in Chemical Change in the following term. Mr Greg told Ms Horton that it was 'not negotiable' that she must finish the objectives and 'fall into line' with the other two teachers who were teaching the unit. He wanted to use a common test to compare all the grade 10 chemistry classes in the school — as a means of identifying the more able students who might be suitable candidates for senior chemistry in grade 11. Mr Greg was firm in his view that Ms Horton needed to cover the objectives of the topic in the current term because these were going to be in the test. On the morning of the 'important stuff' lesson he had informed her that the test was to be held early in the following week:

This morning I went to negotiate with him to see if it was okay if I had the test next Friday. I was going to give it to them at that time. But he said no, that all the classes needed to do it on Monday. So when I checked through the objectives before this morning's lesson, I realised that I needed to teach valencies. I had to do it. I needed to cover that type of work.

Thus Ms Horton's account of events was that she had been under pressure from Mr Greg to quicken the pace of her teaching. At first she resisted, arguing that she needed more time to consolidate her students' understanding of the concepts. However, as the end of the term got nearer, Mr Greg became more insistent that Ms Horton cover the essential objectives of the topic in the time allocated.

DIFFERENT REALITIES: DIFFERENT READINGS

The foregoing descriptions of Ms Horton's grade 10 chemistry class focused on the activities of four members of the classroom and school community. The practices of each member of this community are shaped by a set of constraints determined by 'social structures, power relationships, and the nature of the social practice they are engaged in' (Fairclough, 1992, p. 72). Each member therefore acts independently but also interacts with others to create a constantly evolving classroom community. To understand the workings of the community, we now probe more deeply into forces that shape and constrain the beliefs and behaviours of the members of the community. In the analysis which follows, we provide three different readings of the forces which shape the behaviour of the main characters — Ms Horton, Mr Greg, Karl and Punipa. The readings are chosen from our knowledge of the literature, our experience in the field in this and other cases, and our framing of the data. The three readings are care and responsibility, social reproduction and learning styles.[3]

Care and Responsibility — Ms Horton and Mr Greg

The first reading concerns the ethic of care and the ethic of responsibility (Gilligan, 1982). In the ethic of care, the teacher is motivated by caring, nurturance and a connectedness to others. It is an ethic that is extremely common among women teachers, but not exclusively so (Gilligan, 1982). For many teachers, particularly those in the elementary service, the purposes of personal care take precedence over all others. They use the imagery of themselves as 'rescuers' or 'haven makers' to describe their teaching (Nias, 1989). In many respects, it is a commitment to the ethic of care that attracts many to teaching in the first place (Hargreaves, 1994) and keeps them there in the face of difficult circumstances. By contrast, the ethic of responsibility emphasises improvements to planning and instruction. This ethic focuses on continuous improvement, instructional effectiveness and student achievement. Schools, particularly secondary schools with a subject matter orientation, are commonly organised and justified according to an ethic of responsibility appealing to professional obligations and improving performance (Hargreaves, 1994).

Here, we suggest that Ms Horton operated primarily from an ethic of care. She was the teacher in charge of student welfare at the school, and was popular with her students. She believed she was in touch with students' needs and aspirations. Her teaching style was friendly and respectful. Ms Horton remembered her own time as a student as being a struggle, an experience that reinforced for her the value of school as mechanism for self improvement. She showed concern for individual students such as Karl:

Karl's very clever when they [his peers] let him do his work. He is very interesting because he obviously gets a lot of freedom at home. I think that he is allowed to drink at home and sometimes seems to come in with a hangover. I usually leave him alone if he does. I worry about him because I think he could do a lot better for himself.

For Ms Horton, school science was important, not for its own sake, but as an emancipatory vehicle for students: 'I really feel that science can empower students a lot'. In this school where few students aspired to higher education, she saw science and chemistry as a means to help students break out of their cycle of low academic achievement and self esteem.

The best thing about teaching is teaching science. Each lesson can bring new insight or understanding. Things can lead to discovery, consolidation, interests that I feel no other subject has the opportunity to change new for old, to empower students to make sense of things and allow students to develop their own potential.

Ms Horton began the topic of Chemical Change as she would any other topic. Her teaching was methodical. Although she was teaching out of field, she tried to ensure that concepts were reinforced before moving on to new material. She was concerned that the students had experience with practical work and be exposed to new concepts. She wanted to extend the more able students, but equally she wanted to hold the interest of the low achievers.

Everyone should encounter the low ability students. They have a lot to teach the teacher about science instruction which can enhance teaching at all levels. It is not that they are demanding, just inquiring, things have to make sense in their own terms, from information they hear from their peers and from home. With the lower ability students, it is a task to complete the objective of the lesson as questions are numerous and interesting.

Mr Greg operated from an ethic of responsibility. He had a strong sense of himself as a leader in the school and in the wider science teaching fraternity. He was a long-standing head of department and a former advisory teacher with a history of successful curriculum leadership in his previous school. His job included the promotion and organisation of the subject within the school. Teaching and learning school science were important to Mr Greg. He was mindful that he was responsible for improving the organisation of the science department.

The school is run down; the facilities are run down; I want to get things organised, get the tests organised, get the programs organised, get the staff working together on common goals. Gosh, look at the labs! They haven't even got gas taps on all the benches. Look at these pigeon holes here! These tests go back to the 1960s! It's a huge agenda.

Mr Greg understood the importance of teamwork, accepting the expertise that each teacher had to offer and living with group decisions. He also hoped that staff members such as Ms Horton would be 'up front' about how they felt, and be confident to put forward alternative points of view. However, putting into practice the goal of working together was not always easy. Mr Greg found that sometimes his ideas — his notions of 'responsibility' — were not shared by his colleagues.

I see my role is to get people to work together and feel happy and comfortable. I want them to have some ownership of what is going on. My biggest problem is that I've got a lot of ideas that I want to put into practice. I think, sometimes, I bowl people over in the process of trying to make things happen. When I realise my staff don't share my ideas I have to back off.

One of the most urgent problems for Mr Greg was the absence of a systematic approach for grading students. When he arrived at the school, all junior science classes were organised into mixed ability groups. The result was that by the time students reached grade 10 there was no mechanism for sorting them into ability groups or to determine who would continue with senior school science units. Students needed to cover a specific body of science content knowledge, especially in physics and chemistry, as preparation for senior science. The need to sort students for preparation for senior science was of serious concern because there was no documentation of courses or records of students' past results.

Mr Greg was concerned, too, about the profile of science in the school. He felt the emphasis on building self esteem put too much focus on the lower ability students. The needs of the more able students were being overlooked. The school's poor performance on state-wide university entrance science examinations was evidence to Mr Greg of the lack of attention given to high ability students. As a result, many of the more able students transferred to neighbouring schools that offered prestigious extension programs for talented science students. He wanted to keep as many students as possible in the senior secondary science classes. His aim was to 'correct the balance' and ensure students performed well in the public external examinations at the end of their schooling. Improving students' performance in the public arena would, in turn, raise the perception of science in the school.

Mr Greg had a clear solution to the problem: an assessment program set by one teacher and administered to all grade 10 students under the same conditions at the same time. The results would then be used to identify the advanced students and to separate them into two classes to form an academic stream to groom for senior science. Mr Greg described how the decision was made:

We selected the unit Chemical Change as the unit of work that would be studied by all grade 10 students. We chose this unit because of its particular content: valencies and symbols and chemical equations. It is the first time students come in contact with the explicit content knowledge of chemistry. This unit gives them a solid body of chemistry facts to build future understandings on. It would be the content tested to make decisions about who would and who wouldn't go on into senior chemistry in grade 11.

Ms Horton was one of the grade 10 teachers who was to be affected by this decision and Mr Greg had a lot of respect for what she was trying to achieve with her class:

I think she shows a lot of concern. Here's a person who is on the high ethical ground. Her interactions with students seem to have underlying high principles. She cares a lot; she's courageous. If a student steps out of line, she'll certainly find a way to let the student know that he or she is out of line. So I see a lot of practical wisdom in the way she interacts with students who are different from highly motivated students with middle class goals.

However, when Mr Greg observed Ms Horton's slow progress through Chemical Change, he was concerned that she did understand the importance of the topic in selecting students for further study in senior school. He wanted to use the common test as an organisational device but this would also ensure that Ms Horton covered the essential content of the topic. Ms Horton interpreted Mr Greg's actions as unwanted (and unwarranted) coercion to increase the pace of her teaching. She resented the external pressures to teach to the common test and would have preferred to teach the topic at her own pace.

I was just doing my job and was mostly comfortable in the standards that students reached in my class. Now I know that the standards are set by external agents to ensure that the students are pitted against a rigorous assessment program.

She had difficulty dealing with the pressure exerted by Mr Greg to conform to his timeline for a common test. Not only did she disagree with his philosophy but with the manner in which he exerted pressure on her to conform. It was not simply a case of being uncooperative, but she 'felt that he was questioning [her] professionalism'. Ms Horton had been at the school for four years and was confident that she was doing something right. When he told her how to do her assessments, she found his approach 'authoritarian'.

With the date for the common test looming, Ms Horton did not feel that she could withstand the pressure from Mr Greg. She became concerned about covering the concepts in time for the test. She did not want her students to be at a disadvantage compared with others in the school. As a consequence, she changed her teaching style to ensure that the content was covered. Her teaching became more focused and more directive. In her words:

I knew that I had a task at hand. I was quite focused and quite upset about male domination tactics [of Mr Greg]. And I noticed that I created an environment of being a task master. I hoped that the kids would notice that I was in a different mood. It wasn't the environment I would have chosen to set. I had to do it. I needed to cover that work and I am hoping that the students realise the reasons for my change of mood.

After administering the test, Ms Horton reflected on the effect that this had on her teaching:

I've certainly shown to myself I have to revert back to the traditional style of teaching. In 14 lessons I had to cover Avogadro's number, molar calculations and all the equations. It's most unfortunate because this is a very interesting subject and I think those students would benefit a lot if they were given a chance to understand the subject without the pressure of rushing to cover content just for the test.

For Ms Horton and for Mr Greg, this episode brought into sharp contrast their different ethical positions, influencing ways of teaching, organising and relating to one another. These differences may well have gender at their heart.[4] For Ms Horton, caring for her students was more important than selecting students for further study in science. She was more focused on student understanding and enjoyment than the subject matter *per se*. Mr Greg's eye was on establishing authority and excellence in the teaching of science.[5] He was aware that Ms Horton is a caring teacher but this was not central to his plans for the department. What Mr Greg interpreted as Ms Horton's stubborn individualism, Ms Horton saw as an expression of her individuality.[6] Mr Greg's attempts to achieve 'consensus' and 'teamwork' around his idea of the common test were received by Ms Horton as 'male domination tactics'.[7] Ms Horton's unsuccessful rearguard action to preserve her independence left her and Mr Greg feeling unhappy and dissatisfied.

Learning Styles — Punipa

A second reading employs the theory of preferred learning styles. This theory proposes that different learners perceive, interact with, and respond to their learning environment in different ways and therefore prefer some environments over others. While there is debate about the origins (and existence) of individual learning styles (Guild, 1994), there is growing popularity for the view that learning styles are culturally derived (Biggs, 1991). It has been noted, for example, that Asian students studying in Australia place a particularly high value on academic achievement and prefer classrooms where the content is presented in a highly structured and authoritative fashion (Ballard, 1989). Such students are said to work in a 'knowledge conservation' mode, relying on memorisation and imitation with an emphasis on summarising, describing and reproducing information accurately (Ballard, 1989). While these generalisations may simply be another example of a Western view that things Eastern are different — what Said (1995) called 'orientalism' — they may provide a useful framework for examining the school experiences of Punipa.

Punipa and her family placed a high value on school. Her parents, who emigrated from Cambodia a few years ago, wanted Punipa to have more opportunities than they enjoyed. They said to her 'go by the rules. Listen to the teachers and study hard'. Punipa understood the importance of school but struggled to adjust to the culture of the school and had few close friends.

Sometimes, people call me names and I just try and ignore them. Saying bad things back to them won't do any good, so I just try and ignore it. Sometimes I feel like running away and locking myself up somewhere. But I don't think that is the answer so I just, like, continue with life. Whatever happens, for better or for worse.

Punipa carried this philosophy into the classroom. Over the years, her grades improved to the point where she was one of the higher achieving students in her year group. She was aware of her place in the academic 'pecking order' of the class and constantly set goals for herself to do better:

I seem to be competitive with the other students but really I'm competing with myself. Each year I set new goals. One of my goals is to beat my last year's marks. I pressure myself to do better than the year before. Sometimes, I want to reach my goal so much, I even come to school when I don't even feel well. I'd do almost anything to get good grades especially for maths and science.

Although Punipa was very competitive, she was reluctant to reveal her ideas in a public way in case she was embarrassed in front of her classmates. She did not usually put her hand up because she was 'scared of making a fool of myself for saying the wrong answer'. She did not like it when the teacher asked her questions and when people looked at her when she answered questions. A large part of Punipa's shyness was related to her concern to 'get it right' in class. She was more comfortable in those situations where the subject matter was presented in a logical fashion and the answers clear cut. This is one of the reasons why science was not one of her favourite subjects:

Sometimes I think science is really hard and when you do experiments you have to try and get the results accurately. I don't like that. I'm scared I'll make a mistake or something. Like I'm scared I'll make a mistake or something but Ms Horton says there are no right or wrong answers. So I don't know what I should do.

Punipa's concern to 'get the results accurately' was evident during the acid/metal experiment — she tried to follow the instructions but struggled to get the pop test to work and, like others in the class, had difficulty understanding the point of the exercise. When she did the experiment, she thought it was 'very frustrating when [she] couldn't get it to work'. Even when she read the instructions twice, she still did not have success. Although it wasn't immediately obvious, Punipa's frustration at not being able to find the right answer was revealed when she wrote to Ms Horton about the follow-up assignment on acids and bases.

Dear Ms Horton
Hi. I'm writing to you about my assignment. My assignment may not be the kind of work you expected, but I tried hard; I really did, honest to God. You will be disappointed when you see that I did one base and no diagram. I couldn't find enough information on my base and one of my acids. Miss, you have got to give me a good mark. Anyway, have a good look through and see what you think.
Thank you
Punipa

In the lesson on neutralisation, I saw a different Punipa. She seemed more comfortable with the style and pace of the lesson. She copied the notes from the overhead and the blackboard and worked steadily through the worksheet referring to her textbook as required. The order, clarity and predictability of this lesson seemed more to Punipa's liking than the open-endedness of the previous experience. In her words: 'I like it when [Ms Horton] is in charge of the whole thing and whatever you're supposed to do. She is the one who sort of takes control. She is the one who helps you and all that'.

Punipa's school experience can be related to her ethnicity. Her competitiveness and determination in the face of racial taunts is consistent with a possible cultural bias for attributing her success and failure to effort rather than luck or ability (Stevenson & Lee, 1990). Her fear of embarrassment prevented her from participating in public classroom discussions and she was uncomfortable when her teacher says that 'there are no right or wrong answers'. Punipa, was more comfortable in those lessons where the teacher was 'in charge of the whole thing'. Punipa's poignant note to Ms Horton illustrates her anxiety about achieving good marks and retaining her place in the academic 'pecking order' of the class.

Social Reproduction — Karl

The third reading concerns social reproduction or the re-generation of existing social inequalities. The focus here is on how the relationship between the dominant culture of the school and the culture of the home and the playground leads many working class children to working class jobs. Paul Willis (1977), for example, documents how working class students in secondary schools in England constantly struggle against the forms of symbolic violence inflicted upon them. Willis proposes that working class students have a cultural preference for things rather than words and for manual rather than mental labour. Thus, unwittingly, they participate in a process of self-induction into the labour process leading to the re-generation of working class culture. Thus, cultural forms, 'residing at the very bottom of our brains', work in tandem with the dominant ideology of the school to help re-generate our social circumstances (Apple, 1990, p. 153).

There is some evidence of this process of social re-generation in the story of Karl whose family migrated from New Zealand 15 years previously. Karl's mother worked as a clothes sorter and his father a bus driver. While his mother was the one who had most contact with the school, Karl's father also talked to him about school. According to Karl:

He'll ask, 'What happened in school today? Did you go on work experience?' And he'll ask me what my friends are doing. My father says, 'Get a good job where there's money and do well at school'. And I might get to work with one of the mechanics for work experience and work on motors.

Like Punipa, Karl had difficulty finding acceptance in the school environment although he formed some firm friendships among his Maori peer group. His membership of the peer group was very important to him and influenced his behaviour in class and around the school.

In most of my classes I do my work. In other classes I listen to my mates. They just talk and I just join in with them. Then I get into trouble and stay in after school. Most times when I'm in trouble it's from my mates, hanging around my mates and that. I get into mischief sometimes.

Karl understood that science and school was important for his future but was not clear how he could achieve success. He wasn't sure what it meant to be right in science. School science was a confusing world — in his words 'mixed up' — beyond both his capacity and the world of his friends.

I think science is important because you need an education to get a job. My mates don't worry about school. They will most probably get a job cooking chickens or something.

Karl enjoyed Ms Horton's friendly approach to him — 'She sort of smiles and says hello and stuff like that' — and the way that she tried to reward people for paying attention in class:

If we are paying attention or working she will give us 30 house points. Then we get 30 minutes off school. She lets us sit in class and listen to music or other stuff. It helps get you through the day otherwise you get a bit grumpy.

Some days, Karl said, his friends 'come to school and don't worry about school and then the next day they might be real good at a subject and help you'. Balancing the competing demands of the two worlds of 'paying attention' and 'mucking about' was difficult for Karl. The pop test lesson illustrates Karl's problem. Although he tried to answer a question, he was easily discouraged from entering the mainstream of the lesson. It was easier and more acceptable to engage in some non-conformist behaviour with friends than to follow a confusing set of instructions.

Similarly, with the lesson on neutralisation, Karl made some attempts to keep up with the material but gave up after a short time and spent the rest of the time alternating between talking quietly to his friends, watching for other interesting happenings in the room and laughing loudly. In this kind of lesson 'sometimes or most times we get bored with it. You're just writing everything down from the board, and books and that, hardly doing any experiments'.

In many ways, Karl and his friends were rejecting the forms of individualism, conformity and academic credentials which characterise schools and science classrooms. Although Karl's behaviour may have been moderated by his respect for Ms Horton, these messages of rejection are still clear. They develop within modes of language, dress, habit and styles of behaviour that demonstrate opposition to the dominant ideology. For Karl, science was not really about science but about negotiating a way of being as an adolescent boy. His alternative behaviours or 'getting into mischief' were experienced as true learning, affirmation, appropriation and a form of resistance. Paradoxically, these forms of resistance to the dominant culture actually contribute to social reproduction for Karl. Setting their sights on 'cooking chickens' or 'working on motors' is a form of resistance which ultimately will lead working class boys back to working class circumstances. Unaware of what is happening to them — and of their own agency in the process — Karl and his alienated working class friends, may well 'graduate' from school to the working class labour market.

CONCLUSION

This chapter is about school cultures and the multiple realities of science classrooms. The central narrative describes Ms Horton's approach to teaching the grade 10 topic of Chemical Change. As she picked her way through the topic, Ms Horton did not recognise the importance placed on this topic in selecting students for further study in chemistry. It was Ms Horton's view that teaching had a 'moral purpose', that she had an obligation to include and involve all students in science learning, and a right to determine her own direction during the topic. She was well liked by her students. However, in spite of her aims, she did not manage to engage Punipa and Karl. Moreover, she left Punipa feeling confused and concerned about her grades. While Mr Greg was sympathetic with

Ms Horton's approach, he had a different understanding of the importance of the topic Chemical Change — a view which seemed more in tune with Punipa's aims. Karl's world seemed to be different again from either his teacher's or the head of science. He was more concerned with his place in the peer group than the means of achieving success in science.

Several questions arise from this study of the multiple realities of one science classroom. Which pedagogies are most appropriate for students of different ethnic groupings? How important is the culture of selection and sorting in science education? Can science teachers afford to ignore this culture in place of more inclusive strategies? Should science teaching have a moral purpose or is science worth teaching for its own sake? What is the influence of peer grouping on student engagement? What role should school leaders play in supervising science teachers? We also wonder about our own positions with respect to these questions. While sympathetic with Ms Horton's aim to create a classroom which is more equitable and more empowering, we wonder how well she is serving her students. Perhaps there is a lesson in Punipa's plea for predictable forms of science instruction. If we are seeking a moral purpose in science teaching, there seems to be a lot more to be understood about school science, ethnicity and the social mobility of students. We can identify with Mr Greg's position but wonder about his method of insisting on a common test to force Ms Horton to change her method of instruction. Finally, we worry about Karl and wonder what kind — indeed if any kind — of science instruction would help him in his struggle to find himself.

Contexts clearly matter in science teaching. The teacher's classroom is embedded within the subject department, which is embedded in turn within the school system, the parental community, and the social class structure (McLaughlin, 1993). These interpretations of context help us understand factors that shape the classroom of Ms Horton. The two students, Karl and Punipa, had different realities from each other and from their teacher's realities, and Ms Horton's realities were different again from her department head's. The experiences and aspirations of these four people did not always match, and often conflicted, with each other and with prevailing views about how science classrooms should be organised.

When we look at science classrooms from different perspectives, the troubling thing is that, in one form or another, they are all true. It is troubling because the teacher, who sits in the middle of these multiple realities, can only attend to the one that is true for her. Moreover, as she rushes headlong through the classroom experience, it is difficult to access these alternative readings.[8] Ms Horton may, or may not, be aware of these alternative readings.[9] In any event, she cannot easily balance the simultaneous and often competing demands of teaching in a caring way, attending to the powerful and pressing concerns of Mr Greg and dealing with the complexity of Karl and Punipa's problems. As researchers we can show how complex these matters are but Ms Horton is faced with these dilemmas in the classroom.

This chapter differs from the previous chapters in that it presents a collage of realities of science classrooms, with implications for reviewing and redirecting some of our approaches to understanding the experiences, knowledge and voices of the participants in the education process. Many of the current descriptions of science classrooms are constructed using voices which have been particularly strident, particularly those of researchers and teachers. Other voices — such as those of students from particular cultural groups — have been less evident. All are framed by their own particular experience of schools. They are necessarily partial. Understanding and bringing together these multiple voices is a complex business. However, it is important to create the kind of classroom and research contexts to allow other voices in science education to be recognised and heard.

NOTES

1. Some of the material for this chapter is drawn from earlier papers by John Wallace, and by Helen Wildy and John Wallace presented at the 1995 annual meeting of the American Educational Research Association, San Francisco.

2. This view of research as a process of elaborating and juxtaposing multiple and sometimes oppositional perspectives has been described variously as a bricolage (Lincoln & Denzin, 1994), contradictorally united texts (Clifford, 1988), productive ambiguity (Eisner, 1997), paratactic arrangements (Jackson, 1995) and dialogically constructed texts (van Manen, 1990).

3. We provide these three readings to illustrate the multiplicity of knowledges present in the classroom. However, in doing so, we acknowledge Ellsworth's (1992) point that these knowledges are contradictory, partial and irreducible. In practice, they cannot simply be made to 'make sense', or be known in terms of single theoretical frameworks.

4. The ethic of care is often characterised as feminine because it arises more naturally out of women's, rather than men's, experiences (Noddings, 1984).

5. Authors such as Siskin (1994) propose that the context of teaching science can lead to a predisposition to order and hierarchy on the part of science teachers.

6. See Hargreaves' (1994) discussion of the difference between teacher individualism and individuality.

7. McLaughlin's (1993) study of the contexts of teaching indicates that teacher collaboration can take on quite different meanings depending on whether the focus is on program and academics or the needs and demands of students.

8. Other authors, such as Paley (1989) and Barton (1998), have written extensively about the difficulties faced by teachers (and students) as they attempt to deal with each other's cultural frames.

9. Ellsworth (1992) argues that the experiences of students can never be fully understood by the teacher. These experiences are unknown and unknowable.

SECTION III

REFLECTION

6. CASES AND COMMENTARIES: GERALD

In the previous chapter, we provided several alternative readings of events which took place in a single grade 10 chemistry class. We argued that each reading represents a different reality for the characters involved in those events. We pointed out the difficulty faced by the teacher, Ms Horton, in accessing and attending to those realities which exist simultaneously to her own. We believe that Ms Horton's difficulties are no different from those of many other teachers faced with the constancy of the act of teaching and limited opportunities to stand back and reflect on the challenges of teaching alone or in the company of colleagues.

In our recent work, we have been exploring this issue of multiple realities and teacher reflection through the use of teacher-written narrative case studies (Louden & Wallace, 1996). Some of these cases were collected in specially convened writing workshops where teachers worked in groups of three or four taking turns to describe a recent successful lesson. While each teacher told his or her story, one teacher listened to the story with the goal of asking questions which helped the story-teller elaborate the details, and another teacher acted as a recorder. Subsequently, the teachers discussed the story, retold it to the teachers present at the workshop, and then prepared a written version of the story. We noticed during these workshops that many teachers showed great satisfaction in listening to their colleagues and telling their own stories about teaching. Moreover, teachers seemed to find the act of reading and interpreting others' stories illuminated their own practice.

Following these early experiences, we refined a way of assisting teachers with their case writing and reflection. We often provide teachers with several examples of narrative cases and a description of the qualities of a case. Paraphrasing Lee Shulman (1992) we draw their attention to several of the features of the case genre — that a case is a narrative; that it is particular, specific to the setting and locally situated; and that it reveals the human condition including agency, intention, motives, frustrations, actions and faults. We ask teachers to focus on a recent or remembered moment in their teaching of some content in a science classroom, to prepare a short narrative case, to give it an appropriate title, and then to assemble a set of commentaries on the case. One of the cases prepared in this way appears below. In it we hear the voice of Gerald, a science teacher.

'GALILEO REVISITED' BY GERALD

Physics has always been of interest to me, but understanding the basic ideas is not one of my strong points. I had taught grade 10 physics sporadically over the last couple of years, and, as a teacher of biology, had always found it tantalisingly interesting yet frustratingly boring. I am sure it came across this way to my grade 10 students. I could never get a total grip on the subject. It seemed to me to be a collection of vaguely connected concepts swamped by inexplicable formulae. As my grasp of basic ideas such as light or the laws of motion was both tenuous and superficial, I could not teach the subject with much confidence. Never was this more apparent than when it came to physics practicals. All the trolleys, ticker timers and light boxes meant little to me — they seemed to attempt to illustrate a small aspect of the theory but never really illuminate it. How could I challenge my students' understanding of the subject if my own had such a feeble basis? This year was different. Close contact with a dedicated physics teacher and access to a great text book (without too many equations) has brought greater understanding for me. I actually looked forward to teaching the topic. However, would my new-found interest lead to better understanding for my students? Would they be able to learn something of significance and relevance? Let me recount one of my experiences this year and you be the judge.

Initially, the unit started with Newton's First Law of Motion — objects at rest will stay at rest, and objects moving at a constant speed or velocity will stay at that speed unless acted on by a force. Understanding of this law depended on an understanding of the term 'inertia' as well as the effects of friction and air resistance. This particular day I ventured into the grade 10 class with the aim of trying to show as many examples of this Law as I could. These included the following examples:

- if you are travelling in a go-cart when it hits a log, the cart will stop but you keep moving at a constant velocity until stopped by a force (namely the ground);

- it is important that children are properly restrained in the back seat of a moving car — if the car stops suddenly, their inertia will carry them through the front window; and
- if you are carrying a large bowl of water and have to stop suddenly, there is a good chance that the water will spill over the side, because it is continuing to move at a constant velocity even though you have stopped the bowl.

So it went on. I was not sure if I was making an impact on them until we started discussing the role of air resistance in stopping objects from travelling at a constant velocity. Now, the class was not even sure what the air was made up of, let alone how air particles might act as a force to slow objects down. I gave the example of riding a bike into the wind. 'What is it in the wind that is slowing the bike-rider down?' I asked. I then offered the idea that the molecules of gases in the air together exerted enough force to be able to do this and was greeted by nodding heads, but blank looks.

Briefly, I then talked about the experiment Galileo was purported to have done on the Tower of Pisa — that is, drop two objects of similar size but different weight. I said we could illustrate this in the classroom to help us understand the effects of air resistance. I held up a flat, wide folder and a piece of paper around the same dimensions and I asked the class to predict what would hit the surface of the desk first. Most students said that the folder would, because it had greater weight. I dropped both and sure enough the folder hit the desk first. I then put the piece of paper on top of the folder and again asked the class how quickly both objects would fall. Again, most said the folder would reach the ground much more quickly, because of its greater weight. I dropped both and, counter-intuitively, the paper did not float off the top of the folder, but fell with it at the same speed. They thought I was tricking them — that I had stuck it on the folder in some way. So, for my *coup de grace*, I scrunched up the paper into a ball, held it next to the folder and asked for another prediction. They still said the folder would be first to hit the surface. I dropped both and, lo!, the paper and folder hit the desk simultaneously (well nearly!).

When I asked for explanations for their observations, many students were quite confused. Their own experience would indicate that the more you weigh, the faster you will fall. However, they had just seen that this was not the case. This confusion was good, because it was a confusion that came when currently held beliefs were being challenged and not simply because they did not understand my explanation. It was important that they go through this confused stage because it meant that they might be open to hearing a scientific explanation of the structure of air and how this might exert a force on objects, a force we call air resistance. From the discussion that occurred as some of the students left the class, it appeared that some of the ideas I had presented had hit home, and that some of their intuitive ideas were being challenged. What a difference this lesson had been to ones on the same topic last year!

COMMENTARIES: ALTERNATIVE READINGS

One of the mysteries of preparing such case study accounts of teachers' work is that they always invite the 'so what' question. What is the point of the story? Or, more politely, what is this case a case of? This is the key question for Gerald and others to answer, if the act of writing the case is to contribute to their learning about teaching. In order to focus their reflection on the cases — and gain some appreciation of some of the other realities of the situation — we asked teachers to collect written commentaries. In this instance, one commentary was prepared by Gerald, another by his head of department, and the third by a student teacher. Finally, Gerald reflected on the three commentaries, and drew some conclusions about the case. A précis of all four commentaries on 'Galileo Revisited' appears below.

Gerald's First Reaction

Three weeks after he wrote the case, Gerald wrote a brief commentary on the case. Looking back on 'Galileo Revisited' he noted that not every lesson since then had been as good as this one:

I am constantly frustrated by my lack of knowledge about this area. Not so much the theory but all the stories, experiences and demonstrations that are so crucial in helping students understand this subject. This story really illustrates that well, and I can remember the buzz in the laboratory with students leaning over to watch the folder and paper fall, arguing amongst each other about their predictions, and asking more questions about what they were observing.

For Gerald, then, the case was a case of success worth celebrating, one of the few occasions when his physics lessons met his own standards of student involvement in learning. When Gerald showed his story to his head of department, Arthur, he received a rather different reading.

Head of Department's Reaction

Arthur began by observing that the story highlighted the problem of non-specialists teaching physics, and suggested a rearrangement of the grade 10 timetable to ensure that it did not happen again next year. Although Arthur acknowledged that the story showed a teacher 'open to learning and listening to others', he was disappointed that Gerald had not come to him for advice. Building on his own substantial specialist experience as a physics teacher, Arthur had many suggestions for activities which would ensure that students interest and excitement was turned into lasting learning:

It was a pity that the teacher left the lesson virtually up in the air. He could have challenged the students to come up with other examples of the effects of air resistance — why do trucks and sports cars have air scoops and bars? How do sailing boats use 'air resistance' to sail against the wind? They could even have designed an experiment to test the best sort of sail. Without such follow-up, I doubt whether the good experience the students had will have any lasting effect.

For Gerald, 'Galileo Revisited' is a case of unusual success in teaching physics; for Arthur it is a case of a potentially interesting lesson which left students 'virtually up in the air'.

Student Teacher's Reaction

Mary, a student teacher, drew a third set of conclusions from the case. What she particularly liked about 'Galileo Revisited' was the insight the story gave her into 'how a teacher really works in a classroom'. She was surprised at how honest Gerald had been, and recognised the same feelings of frustration in her own work on teaching practice. She particularly liked the way in which the story communicated the atmosphere of the class and the values of the teacher:

The story gave me a real feel for the classroom — the questions, the discussion and the responses. You could almost sense the atmosphere . . . I can also get an idea of the things that are important to this teacher — the frustration of not fully understanding the subject in previous years; the excitement that results when students get enthusiastic about a topic and want to think about the concepts involved; the importance of being confident when teaching a topic.

Gerald's Second Reaction

After preparing his own first response to the case and collecting two other commentaries, Gerald was asked to reflect again on the experience of writing the case and reading the commentaries. He liked the idea of writing about physics in a narrative form, and thought that it might be useful to show students 'as a means of getting them to reflect on themselves as learners and students in a classroom'. Gerald was also pleased with the response from his physics department head. Despite his embarrassment at not seeking expert advice earlier, he thought that Arthur's comments contained some useful and helpful suggestions that he 'would like to follow up'. Beyond the specific advice he received, Gerald also valued the opportunities to open up professional conversations with colleagues:

I think that a certain amount of professional pride on my part prevented me from asking him for help. However, I think that this will improve in the future. This exercise then has been another way of getting to know teachers beyond the superficial lunchtime chat sessions and it shows them something of my character and methods.

Finally, Gerald commented on the value of collecting commentaries on his case study. Reading over the case and the commentaries helped him 'reflect on the lesson' and 'see what really happened'. He had been successful in sparking interest, but was less successful in building on the interest. Next time, he resolved, he would try to capitalise on the interest at the time and not 'wait until the next lesson'.

READING CASES AND COMMENTARIES

Preparation of the cases and commentaries led Gerald through a cycle of reflection on his successes and failures in teaching physics. Through his selection and shaping of events into the story 'Galileo Revisited', he confronted an enduring problem and celebrated his success in overcoming the routine of definitions, formulae and half-understood practicals which dominated his experience of physics teaching. The comments provided by his head of department, Arthur, added to his store of examples to use in teaching the topic of inertia, and encouraged him to think more about using students' enthusiasm to build lasting learning about physics. Mary's and Arthur's recognition of the 'character and methods' displayed in his story helped him overcome the professional pride that sometimes prevents teachers from sharing their uncertainties with colleagues. His reflection on this case made a contribution to his practical knowledge of teaching and, perhaps more importantly, it allowed him to move cautiously beyond his own classroom and open the professional dialogue with some of his colleagues.

Notwithstanding Gerald's satisfaction at the responses he received, as teacher educators we had been hoping for more from the commentaries offered by teacher colleagues. They seemed generally too thin or superficial — invariably polite ('The math lesson sounds like good fun. It makes me wish I could be in Marion's class'), but sometimes negatively critical ('Henry's candor is engaging, but an inadequate substitute for mastery of subject or teaching principles'). These kinds of responses are similar in character to those provided by peers in the context of peer supervision (Wallace, 1998). We speculate that colleagues are like invited guests when asked to observe and comment on a fellow teacher's performance, involving certain behavioural expectations. These behaviours operate as protocols for both parties, for example 'invite your friends', 'present your best effort', 'don't be longwinded', and 'avoid criticism'. Indeed, we also found ourselves reluctant to criticise Gerald's performance. Perhaps this is because we have been researching the consequences for teachers of gaps in content knowledge and regard this as one of the endemic problems of science education (Parker, Wallace, & Fraser, 1993). We were reluctant to criticise Gerald for his intelligent and energetic attempt to overcome a problem that he shares with many science teachers. Secondly, we were more inclined to applaud than condemn Gerald for 'entertaining' his students in physics. Our sense of contemporary school physics teaching is that the discipline is attempting to move beyond Arthur's classically gendered examples of trucks and racing cars, and towards contexts that make physics more meaningful to a wide range of students (Wildy & Wallace, 1994). Thirdly, we wanted to support Gerald for his view of students' learning in science. When he says: 'I could never get a total grip on the subject. It seemed to me to be a collection of vaguely connected concepts swamped by inexplicable formulae', he is articulating a familiar critique of traditional physics teaching. In sharp contrast with the 'number plugging' algorithmic approach which has dominated physics teaching for many years, Gerald recognises his students' confusion as providing a teachable moment when he says that 'this confusion was good, because it was a confusion that comes when currently held beliefs were being challenged and not simply because they did not understand my explanation'.

On the other hand, when we discussed Gerald's case with our own academic colleagues expert in physics teaching, we had a sense that for some readers the teacher's confusion was more salient than the students' confusion. A commentary prepared by Bevan McGinnis picked up this point. After acknowledging that he might have had similar problems teaching biology, Bevan went on to say:

I found it not surprising at all that his students were confused. If they did not even understand what air was made up of, then linking inertia (a new concept) with air resistance without explaining that it is friction, would confuse them. In fact air resistance as a force is a little more complex than it first appears. It is a reaction force, and hence is better left to a discussion of Newton's Third Law. It is also a force which acts in concert with gravity on falling bodies. A body will accelerate as long as there is a force acting on it. In the case of a falling body, this force is gravity, but not only gravity is acting on it. The air acts on it also, slowing its descent, but as the body travels faster, the air resistance increases until finally it equals gravity and the body stops accelerating. The speed it attains is called terminal velocity.

The famous, if apocryphal, experiment that Galileo performed had nothing at all to do with inertia or indeed air resistance. It could not be adequately explained until Newton formulated his laws of gravitation. It was dealing with the observation that the force of gravity is not dependant on the mass of the body. But the speed a body falls at is neither dependent on merely gravity nor on inertia, rather called air resistance. In fact what Galileo is purported to have proved was indeed wrong, objects that are different sizes do *not* fall at the same speed. They fall with the same force of gravity, but that is not the only force acting. This is dealt with in the second law of motion. Had he performed his experiment of the moon, where there is no atmosphere, it would have worked.

I have a bad feeling that these students may end up with the idea that inertia is somehow linked with air resistance, which it is not. Inertia is a property of a body and is dependent on its mass. All the other things Gerald spoke of were demonstrations of the interactions of forces, and as such were good, but not relevant to inertia. Also the folder and the paper demonstration could lead to a confusion of the concepts of mass and density. It also had nothing at all to do with inertia.

I have no wish to sound disparaging of the efforts that Gerald made, but his physics was confused and the demonstrations that he used were mixtures of different concepts — inertia, friction, the interaction of forces, air resistance and its dependence on surface area, and the Newton's second and third laws of motion as well as his law of gravitation. Discrepant events are an excellent way of stimulating thought, but when they are used to start a new concept, it is important that they are adequately explained.

For this academic reader, with a deep understanding of physics content and pedagogical content knowledge, Gerald's constructivist reading of his story is not adequate. Bevan is not willing to privilege the confusion which accompanies students' exploration of discrepant events, if the discrepant event leaves students unclear about the difference between fundamental physics concepts such as inertia and air resistance.

Other Cases, Other Readings

We have chosen to focus on Gerald's case material because it illustrates a common and fundamental issue: how hard it is to teach well when we lack a deep understanding of the subject content. Which interpretation of 'Galileo Revisited' to prefer, we leave to readers. Neither Gerald's commentary, the commentaries of his colleagues, nor the commentaries of our academic colleagues can exhaust the meaning of the story. The point we want to make is about opening up the possibilities for learning through teachers' stories, not closing them down by providing an authoritative interpretation.

Many of the stories written by the other teachers we worked with were equally powerful. As it was with Gerald's story, however, the commentaries teachers collected seemed less rich than their cases. Sometimes responses to teachers' commentaries ranged from patronising to hostile. One teacher, Scott, for example, said 'I did not need a reminder about the issues of contract deadlines and documenting student progress'.

Othertimes, teachers thought that the commentaries produced by their colleagues were too coy. Another teacher, Michelle, for example, observed that 'the teacher who knew my identity was intent on voicing her approval of my approach to teaching science'. For Scott, feedback from his experienced colleague was 'disappointing' because he questioned the suitability of Scott's student-centred teaching methods. A third teacher, Anna, expressed 'surprise' at the reactions of two colleagues, once because of what Anna regarded as an inappropriate focus on assessment, and once because she had not expected so personal and searching a response from her colleague. Regardless of the quality of these commentaries, we feel that thoughtful responses to stories are important if case writing is to lead the author to restorying (Clandinin & Connelly, 1992, 1995) and hence changes in practice.[1] Among the most detailed and helpful commentaries our teachers received was one from Scott's colleague, Bob who was an experienced university science educator. Responding to Scott's story, Bob provided a conceptual context for the events Scott described and identified several issues for further consideration:

> Bob's feedback was welcome. He has left me with the challenge of understanding why I chose particular actions. This will require more reflection on my part. . . . In the short term, Bob's comments have encouraged me to continue with the intended expansion of methodologies employed by science teachers in this school.

Scott was fortunate in having a collaborative relationship — combining the closeness of a friend with the distance between teacher and academic — which allowed Bob to encourage Scott and to challenge him to understand the decisions he had made. Such collaboration between teachers is never easy to achieve. At the least, it requires a measure of trust, a balance between professional pride and professional humility, careful negotiation of differences in power which may exist, and acknowledgement that each party can expect to learn from the other. These qualities of collaboration are discussed more fully in Chapter 8. Gerald's commentaries do not encourage us to believe that he was able to find suitable collaborative partners in his school, though his second reaction indicated that he thought the act of sharing his story would help him move 'beyond the superficial lunchtime chat sessions'.

CONCLUSIONS

We share with many others an interest in exploring the use of case studies and narratives (Carter, 1993; Sykes & Bird, 1992). In particular, we have been interested in encouraging a disposition towards reflective practice. Our analysis of the different forms and interests of teacher reflection are the subject of Chapter 7. The idea that reflection can set teachers free from routine, habitual patterns of thinking and action is at least as old as John Dewey's *How We Think,* published originally in 1910. Equally old is the criticism that members of the teaching profession lack a reflective disposition. Writing more than sixty years ago, Willard Waller noted a 'peculiar blight which affects the teacher mind, which creeps over it gradually, and, possessing it bit by bit, devours its creative resources' (p. 391). He attributes teachers' lack of flexibility and adaptability to an over-adjustment to the 'simple, changeless rhythms' of teaching:

The initial pleasure of the teacher in the teaching role endures throughout the period in which teaching habits are being formed. During this period the work of teaching furnishes nearly the entire content of consciousness, and is attended to almost exclusively. As habits recede from the consciousness with increasing perfection, satisfaction fades, and discontent grows. This . . . analysis of teacher discontent points towards routinization as one of its major sources. (Waller, 1932, p. 436)

Nearly 70 years on, Waller is not alone in this view. We share his concerns, noting that even when conditions are 'right', and when teachers have time and a safe space, they often don't know how to think about their work in productive ways. Like Ms Horton, discussed in Chapter 5, they are caught in their own reality and cannot move around, or through it, to appreciate the classroom in other ways. Often, they remain at the level of reporting what occurred, rather than moving to a level of analysing, which requires particular reflective skills. Nonetheless, we think that preparation of narrative cases and commentaries is one useful way of helping teachers hold off the descent into unreflective and unhelpful routines. Cases provide a means of recording the wisdom of the profession in a way that is accessible to teachers. The act of writing provides a way of discovering sequence in experiences and understanding cause and effect relationships in past events (Welty, 1984). Thoughtful commentaries can open up the possibility of other readings or realities beyond those of the classroom teacher. The value of cases lies in their potential to represent the messy world of practice, a world where often neither the problems nor the solutions are clear. Often this potential is best realised in collegial settings where there are opportunities for reflection, discussion and joint problem solving.

NOTES

1. Clandinin and Connelly (1994) make the point that it is the retelling of the original story that allows for growth and change.

7. INTERESTS AND FORMS: JOHANNA

For individual teachers, such as those described in the preceding chapters of this book, the process of professional development is a process of gradual change in understanding and action. Over time, content knowledge grows, a repertoire of patterns of teaching emerges, and hopes and dreams for the future change. Throughout this career-long process of development, teachers reflect on their understanding of their work and on the relative success of the kinds of classroom action they have learned to take. They look back on old ways of doing and thinking, settle on practices which seem to work, and look forward to new ways of resolving the challenges of teaching.

The purpose of this chapter is to explore the range of ways in which reflection contributes to the development of teachers' learning. Like the case study material described in Chapter 2, this chapter draws on a year-long case study with a teacher we called Johanna[1]. Using some of the many stories Johanna and I told each other about our collaborative teaching, the chapter builds a conceptual account of teachers' reflection. Briefly, this conceptual framework includes two complementary dimensions, dimensions which I call the *interests* and *forms* of reflection. These two dimensions are developed from the work of Habermas (1971) and Schön (1983, 1987) respectively. The term *interests* refers to the goal or end-in-view of an act of reflection: is the goal of reflection fidelity to some theory or practice; or deeper and clearer personal understanding; or professional problem solving; or critique of the conditions of professional action? *Forms* refers to the characteristics of the act: is it a matter of introspection, of thinking and feeling; of replaying or rehearsing professional action; of systematic inquiry into action; or of spontaneous action? These dimensions, it will be argued, are different and complementary. A particular act of reflection thus has an interest and a form, and in principle all reflective acts may be described in terms of both dimensions. For example, one might use a form of reflection such as inquiry into action with the end in view of critique, or problem solving, or personal growth, or technical fidelity to theory. Equally, one might serve a critical interest by introspection, or by rehearsing a range of options, or by a process of inquiry, or through some spontaneous discovery made in the midst of professional action. In the discussion which follows, I first describe the four interests and then connect the four interests with the four forms of reflection.

REFLECTION AND INTERESTS

One strand of writing on reflection (Carr & Kemmis, 1986; Grundy, 1987; Tripp, 1993) has drawn on Jurgen Habermas' theory of 'knowledge-constitutive interests'. Habermas (1971) distinguishes among the interests of the empirical-analytic sciences, the hermeneutic-historical sciences and the critical sciences. Habermas associates each of the forms of inquiry with a cognitive interest: empirical-analytic inquiry with technical control by discovering rule-like regularities in an objective world; historical-hermeneutic sciences with practical control through understanding and communication; and critical sciences with emancipation through critical reflection on the conditions of social life (1971, pp. 301–317). This chapter builds on Habermas' framework, following the *technical* and *critical* interests he distinguished and separating the practical interest into two categories, a *personal* interest and a *problem solving* interest. Both of these latter interests share Habermas' sense that the historical-hermeneutic sciences serve the interest of practical control, but what I have called the personal interest emphasises the personal meaning of situations and the problem solving interest emphasises resolution of the practical problems encountered in professional work.

Technical Interest

The technical interest, an interest in controlling the world by attending to rule-like regularities, is a powerful force in education. It stands behind quantitative research into effective schools and teachers, competency-based teacher evaluation, and much of the research into curriculum implementation (Wallace & Louden, 1998). Key issues in technical reflection include fidelity of teachers' practice to some set of empirically or theoretically derived models and the development of technical skills of teaching.

In all of the discussions I had with Johanna during a year of collaborative work, there were no clear examples of reflection with such technical interests. In general, Johanna was suspicious of the plans and programs promoted by her school board and unimpressed by the possibility that she would improve her teaching by following prescriptions of people who were no longer involved in classroom teaching. The closest Johanna came to using propositional knowledge during the life of the study was to refer, several times, to her reading of a pair of books by Thomas Gordon, *Teacher Effectiveness Training* (1974) and *Parent Effectiveness Training* (1970). On the day we first talked about the possibility of participation in a collaborative study, she mentioned that she had recently read one of these books and that it had made a substantial impact on her. Much later, when she was planning a curriculum night she considered looking in these books for activities which might help parents understand her approach to problem solving with their children. She had come to these books at a time when she was trying to improve her communication with her daughter, and thought of them as symbolic of the changes she had been making as a teacher in the time just before we met:

I found the books the summer before you came and they had really changed the way I was working with kids. I found that I was having tremendous difficulty here — tremendous success in many areas but also tremendous difficulty . . . I really wanted to be here and I was trying to figure out how I was going to do that.

Johanna had not attended a training program using this approach, and she did not make detailed references to skills or principles she had drawn from these books, but they had come to symbolise something special and important about her philosophy of education and the approach to children she had been trying to bring to Community School. She rarely talked about teaching in terms of skills, preferring to talk in terms of 'tricks' which she might trade with other teachers at, for example, drama conferences.

For instance, working in groups of three I have got in my bag of tricks different things I can have them do with the same information. I can have the kids sitting back to back so that they can't see each other while they are talking, and both doing a running monologue about how they are feeling at the same time. It's another way of approaching the same problem. I can switch and have them do whole group work, and if kids are finding it too hard, too embarrassing, I can take the focus of that and put it somewhere else. . . . You have to have those skills and that's why I go to drama workshops, because people will teach me different things to do with kids, things that I wouldn't have tried that work brilliantly. And once that becomes part of your repertoire, then you can trot that out at any time.

The kind of reflection involved in trading these tricks, of course, barely fits the notion of technical reflection. Her interest was not in fidelity to prior theory or practice, but in expanding her repertoire. And as Johanna pointed out in this same discussion, 'You only take it in as it relates to your own life, anyway'.

Personal Interest

Much more common in Johanna's work was reflection with a personal interest, an interest in connecting experience with her understanding of her own life. Such a personal interest in reflection informs Clandinin and Connelly's (1991) narrative method in teacher education. Narrative, they explain, is 'the study of how humans make meaning of experience by endlessly telling and retelling stories about themselves that both refigure the past and create purposes in the future' (Connelly & Clandinin, 1988, p. 24). Johanna told many stories about herself which explained the biographical connections between her experience and her actions, and which shaped her sense of how she ought to act in the future. She talked, for example, about the biographical roots of her very earnest and serious approach to teaching. As a child at primary school she felt that her job was to be intelligent, helpful and well behaved, and she took life very seriously:

That all came from the fact that I was the child after a Down Syndrome kid, and I think that when I went into that school I was a kid with a purpose. I new exactly what I had to do in life. These were things that were expected of me and I knew really clearly what I had to do. I was really serious. I can remember my Dad . . . saying to me, 'Who's the funny one in the group?' And I said, 'Well, I guess it's me'. And he just laughed. He said, 'You!' as if it was totally impossible that I would be funny. And I thought, 'Oh, I guess I'm not funny'. But there must have been some of the big me — that loves a joke — in the little me. I still do take life very seriously, and I don't think that there's anything I can do to change that. That's who I am.

Similarly, she talked about the connection between her commitment to helping students to become more independent and her own experience of uncertainty at college during the 'druggie times' of the Sixties:

It really affects what I do with kids. My feeling about why that time was so absolutely terrifying for so many kids is that the world didn't make any sense to them in terms of their own independence. They were not strong, they were not solid and I certainly wasn't. I'd never been given practice in decision making. I was looking for some sort of meaning for life, for some sort of truth that I was told was very important, and then handed Catholicism. So, for me that was a terrifying time because, although it was very exciting, you were being blown about by winds that were bigger than you and you could be so badly hurt. People were dying, people were going crazy. It was heady, because you finally had your freedom and you were beginning to do things together as a group, but we were such children.

Her education and upbringing had not helped Johanna learn to make decisions about her own life, she thought, and this had consequences for her actions as a teacher:

You don't have to be a baby at 18 and 19 and 20. But our society made us children and kept us children. It is something I feel is terribly dangerous to children. You have to give them the strength to go out and be able to protect themselves and be independent and make it in the world and not be led by peer pressures into stuff that is going to be destructive. So, that experience of being so totally unprepared led to how I feel I ought to deal with kids now.

For Johanna, one of the effects of personal reflection is that it supports her sense of agency, her sense that she controls her own destiny. From her father — 'who always wanted a boy', she said — she received permission to be the one who decides what she should do, to be the wage earner, to take control rather than be controlled by others. Subsequently, and through a series of changes in country of residence and subject specialty as a teacher, she struggled to construct a safe and comfortable working environment. At Community School, she felt that she achieved the freedom to do what she wanted to do as a teacher:

Right now I have the freedom to really do what I want to do with the kids, which is an unbelievable freedom. Very, very few teachers have that freedom to work in an environment of people that I know and love, and do what I want to do without having too much interference from people who are trying to tell me what I ought to be doing. I am in a wonderful situation at the moment. If I can make it less work for myself and less exhausting, then it is a great position to be in.

Problem Solving Interest

Unlike reflection with a personal interest, which connects biography and experience, the problem solving interest is concerned with resolution of the problems of professional action. This is the interest most fully represented in Schön's (1983, 1987) well known work on reflection. The problems of most concern to Schön are problems which fall outside the established technical knowledge of a profession: cases, for example, which are not 'in the book' (1987, p. 34) and situations which are 'uncertain, unique or conflicted' (1987, p. 35). Such problem solving may take place in informal frame experiments which take place while it is still possible to alter the outcomes of action — Schön's reflection-in-action — or after the event, as in Schön's reflection-on-action. In either case, his interest is primarily with the situations that learners or practitioners already see as problem solving: occasions where people are surprised by what happens and are moved to rethink their professional practice.

There were many examples of reflection with a problem solving interest as Johanna and I taught writing and science together. Some of this reflection was what Schön calls reflection-on-action. We talked, for example, about problems with content, alternative patterns of teaching, ways of pursuing independent learning goals, and the problems we had with particular groups and students. More interesting, and less well documented in the educational literature, are those examples of the reflection with a problem solving interest which Schön calls reflection-in-action. The vignette below, from a writing lesson, provides several such examples. This lesson was part of a long sequence of lessons during which Johanna's classes prepared the text for their own illustrated books.

Vignette 1: 'That worked well'.

About two weeks into the illustrated book project, Johanna had asked students to complete the text of their books and to be ready to hand them in. When she announced to the students in grade 7 that she was coming around the class to check, some of the students attempted to talk her out of collecting the work, claiming that they didn't know about the deadline or shifting the blame to some third party. Johanna was irritated because she had planned to take the scripts home and correct them before students started to do their illustrations. How could she do this if students were not all finished? She took out her mark book to record those who were ready and who were not.

How many people are done? I am going to take this down. A little language mark. [Writing in a column of her mark book.] 'Story by due date'. If you have done what was required for today — no excuses count — if you were done by today, raise your hand.

Just as she realised that only five of the sixteen students present had done as she asked, a student arrived with a message from the school secretary, asking her to check the form which she had prepared for students' interim reports from the school. Johanna seized the opportunity offered by this interruption, connected the mark she had just allocated for meeting the due date to the interim report, and began reading aloud from the interim report form.

Listen up. Interim reports are coming out soon. Here's what the interim report which is going home on Monday is going to say. You could be in the 'A' category: 'This student is making good use of Community School', or you could be in the 'B' category: 'This student is progressing satisfactorily. The check marks indicate areas which the teachers consider require special attention: homework, behaviour, punctuality'. Punctuality! 'C': 'A teacher who places a check mark in one of these boxes does so because he or she feels grave concern for this aspect of the student's learning and would like to discuss the matter: Math skills, reading skills, need for supervision, ability to meet deadlines'. Ability to meet deadlines! Five of you passed test number 1, ability to meet deadlines!

While students were finding their calendars and marking in their weekend homework, I walked across to Johanna, who had been sitting on a stool on the rug in the centre of the room. She turned to me and said:

Johanna	What shall we do?
Bill	I was just going to suggest that one way of saving you some marking time at home would be for the two of us . . .
Johanna	To travel round?
Bill	. . . to read their stuff while the students who are ready continue with their reading.
Johanna	OK, I'll let the five students who are done do their illustrations. [*To class.*] OK, we are going to have two different lessons happen today. The first lesson is for the five people who are ready — and this is what

you would all be doing if you were ready — can start to work in their
sketchbooks on the illustrations for each page.

While Johanna moved around the room helping students with proof-
reading, I sat at one table and worked through several students' stories in
detail. At the end of the lesson, Johanna said, 'That worked well. We can
keep helping them while they do their illustrations. This would be better
than taking it home to check'. I replied, 'Exactly, the point is that it gives
us the chance to teach at the point of error rather than taking it home and
practising our own spelling and punctuation'.

* * *

There are two instances of the form of problem solving reflection
Schön calls reflection-in-action and one of reflection-on-action in this
vignette. First, when Johanna was interrupted by the student carrying
sample report forms just as she realised that only five of her sixteen
students had met the absolute deadline she had set, she seized the moment
and connected the two events. Like Schön's example of a jazz player's
reflection-in-action, she smoothly integrated a new element into her
ongoing performance (Schön, 1987, p. 30). Second, when she realised
the impact this would have on her lesson plan, we had a quick
conversation about what to do next. Like Schön's example of reflection-
in-action as he built his garden gate (Schön, 1987, p. 27), we were
surprised by what happened and invented a new procedure in the midst
of action. Had Johanna been on her own, she might have taken a moment
to have a similar, silent conversation with herself. Third, as the lesson
closed we swapped stories about our responses to this unexpected turn
of events. As we reflected *on* what had happened, Johanna commented
that the split-second decision we made (our reflection-*in*-action) had
worked well, and I observed that this pattern of teaching might help
solve a problem she had been worrying about — finding a way to help
students with their sentence construction.

Critical Interest

Although perhaps the least familiar interest for reflection among teachers,
if the reports of Jackson (1968) and Lortie (1975) are to be believed, the

critical interest is the most comprehensively theorised of the four interests of reflection. The essence of the critical interest in reflection is that it involves questioning taken-for-granted thoughts, feelings and actions. Through such reflection, teachers may confront and perhaps transcend the constraints they otherwise perceive as normal or natural. Critical reflection begins with the assumption that reality is socially constructed and that people can act to influence the conditions in which they find themselves. To this end, critical reflection involves considering who benefits from current practices, how these practices might be changed, and personal or political action to secure changes in the conditions of classroom work.

Vignette 2: Active Learning

It was not always easy for Johanna and me to find time for reflection at Community School. Often we would have lunch together at the bistro across the road from the school and there, freed from the press of students passing through the staff room asking questions and looking for company, we were able to talk about a wider range of issues. We talked at length, for example, about the idea of a good school and about the possibilities of educational change. In one of these discussions, Johanna talked about a school board seminar she and a colleague, Miles, had just attended. The topic was 'active learning', a major local priority in the year of the study. She was enthusiastic about the workshop, partly because she agreed with what the speaker had been saying and partly because he had said it in front of her principal:

Miles' comment was that what he said we all knew fifteen years ago — but he said it to a gathering of every single teacher in this area. He said it in front of the superintendent and our principal. He'd been hired to come and say it, so obviously it was approved of.

What they said was exactly what you and I have been saying, which is that you cannot pour knowledge into a kid. What a kid actually learns will depend on how actively involved in his own education and problem-solving he is and that to delineate the number of minutes [per week] that you have him doing a certain thing doesn't give you any information about what has happened educationally for that child.

So it was a vindication and I looked around to see if [the principal] was listening to that little bit because it is what we have been saying to him.

I was glad that she felt vindicated by what the speaker had said in front of her principal, but I wondered aloud what other people might have made of the session. There was obviously a good fit between Johanna's teaching and the idea that was being promoted. I wondered, however, whether such a session would have been any use to a teacher with a more skills-oriented approach to teaching. Did talking about 'active learning' at a workshop make any difference to such teachers? My question seemed a little pessimistic to Johanna, who knew how much she had changed as a teacher in recent years. She had, she said, spent many years 'trying to impose the way I felt things ought to be on kids, instead of listening to where they were'. Perhaps such sessions could help other people to move along the path towards active learning. My reaction was that such sessions would have no impact on Johanna's own teaching:

What they are ignoring is how deeply connected what you did today is to your history. You are an 'activity' teacher. When you get a subject which lends itself in other people's hands to giving lectures and notes, you turn it into an activity subject. Invariably, you go for the kid's personal difficulty not the subject content.

We realised that we had a fundamental difference of opinion about the value and the possibility of making a particular teaching method compulsory. The difference between us was not in our vision of good schooling but in our sense of the practical and ethical problems of communicating our vision to other people. Johanna, perhaps more committed to the value of active learning than concerned about what it would be like to have someone trying to impose a different teaching method on her, responded with an assertion of the need for change and a practical example of the way she would approach it.

I still feel that there is a way. When you are at teachers' college you do courses which are theoretically supposed to help you understand the human mind. I would like to teach those courses by having the people actually being involved in the technique of active listening. . . . To give you an example, yesterday I was in this drama group with a guy from England. I watched what happened in the group. One of the people in my group was very bossy and I was put off by her approach. I could sense that there were several people in the group who had never done any drama before and were very nervous. Because of what I know about how to listen to people, I was able to get the group working happily together. . . . Now I could teach that to teachers and that would be a very worthwhile thing, because that's what we do — work in groups — so I would say that would be a way of changing people.

* * *

In conversations such as this, one interest in reflection may be dominant without other interests being abandoned. Here, the interest is primarily critical, because it concerns the application of power over teachers and the possibility that well-meaning educational change activities may close down some people's opportunities to teach in ways that are consistent with their own biography and experience. But alongside this interest in the ethical conditions of the use of power is Johanna's personal interest in ways of teaching that are biographically appropriate for her, and my personal interest in the need for plurality in educational change. More than this, the critical interest in the discussion is pursued through the example of a practical problem and the technical skills Johanna used in overcoming the problem.

FORMS OF REFLECTION

These four *interests* of reflection account for the range of reasons teachers might have for reflection, but they do not help elaborate the range of ways in which changes in understanding and action take place. To explore the latter issue more closely, I now turn to the dimension of *forms* of reflection. This set of categories represents a range from reflection as a process of thinking or feeling separated from action to reflection as a process which takes place in the moment of action. Between these extremes stand two categories of reflection which deal with both thought and action.

The reason for arraying forms of reflection across this dimension is that the forms ought to allow for the range between tacit and explicit knowledge. Some of Johanna's knowledge as a teacher is tacit and embodied in her practice, in patterns such as the way she conducts class discussions or conferences, the way she begins and ends lessons, and the way she teaches guitar to whole-class groups. Other parts of Johanna's knowledge are more explicit, such as her content knowledge in music and art, and the hopes and dreams for teaching which impel her to work with students' feelings and to prefer content which is relevant to their interests. Some of her knowledge is not knowledge at all — in the sense of knowledge as justified true belief — but a series of unanswered questions on which she is still working, such as how she can best help students to become more independent and how she can find a comfortable way of teaching science.

Consequently, this dimension ranges between forms of reflection appropriate to the two extremes of tacit and explicit knowledge. At one extreme, reflection may be a conscious process conducted at some distance from the stream of action. This form of reflection, which involves both thinking and feeling, may be called reflection as *introspection*. At the other extreme stands a form of reflection so bound up in the moment of action that there is no conscious awareness of thinking about the action, a form of reflection which may be called reflection as *spontaneity*. The two intermediate categories are reflection as *replay and rehearsal*, the sort of reflection which might involve a teacher thinking or talking about events that have happened or might happen in the future, and reflection as *inquiry*, a form of reflection which involves thinking and acting in a deliberate process of inquiry. Whereas Schön's dichotomy of reflection-in-action and reflection-on-action accounts for this range in two forms of reflection, the remainder of this chapter explores these four forms of reflection and connects the forms of reflection with the four interests already established.

Introspection

Introspection, which involves looking inwards and reconsidering one's thoughts and feelings about some issue, is the closest form of reflection to the ordinary language sense of reflection as contemplation or meditation. In Johanna's case, most of the introspection she shared with me had a personal interest: the relationship between her family of origin and her attitudes to life and work; the relationship between her experience of the 'druggie times' of the Sixties and her determination to help students learn to accept more responsibility for their own decision making; and the effect of learning clarinet as an adult on her attitudes to praise in music teaching. Less often, this introspection led to reflection with a critical intent, such as in the following story about the experience of growing up during the second wave of feminism:

There was an interesting episode in high school with my two crazy friends, who certainly were not the twin-set-and-pearls style. One of them was on the committee where they could elect the Prom Queen, and my name came up. My friend told me about it later. She said, 'I couldn't do it to you Johanna, I just couldn't. I got you out of it'. At the time I had the feeling that it was bizarre that I would want recognition from that world, of being the female symbol of sexuality. I was saved by my friend but I wasn't sure I was saved. I wanted it, yet it would have grated. The same sort of thing continued at college. I didn't know where I belonged. At the fraternity, one of the things they did was hold a huge parade and they would find girls to march along the parade wearing short skirts and looking sexy. I can remember doing that and thinking, 'Nah. This is not where I want to be. This is me on a platter, served up'.

It was very hard for me to find a way of being feminine and sexy without going into that world where women are objects. So for a long time I couldn't pass through a cosmetic department in a big department store without feeling that the women who stood behind the counters would recognise that I was not one of them. I wasn't a type who could do that. Somehow I would like to have been able to be female in that way, whatever that was. I had to find that much later, how to be feminine and not be a possession.

Replay and Rehearsal

Replay and rehearsal is a form of reflection which involves teachers' talk about events that have occurred or the possibility of future actions. As teachers talk to their colleagues (or write) about their work they make sense of surprising classroom events, draw provisional generalisations which may inform their future practice, make plans for action, and affirm their values. This form of reflection is one step closer to action than *introspection*, but still stands at some distance from the deliberate movement between action and reflection which characterises *inquiry*. Teachers work in conditions characterised by immediacy, multidimensionality, simultaneity and unpredictability (Huberman, 1983). Only rarely is it possible for teachers to think or talk about the meaning of their experience or their immediate plans while they are still in what Schön calls 'the action present' (1983, p. 62). More often, teachers are too fully immersed in what they are doing, too busy to consciously reflect on what they are doing while there is still time to make a difference to the situation at hand.

Unlike the 'virtual worlds' which Schön describes in the profession of architecture (1987, pp. 75–78), it is not easy to construct realistic models of teaching where teachers may practice and refine their actions. Classrooms are usually too busy to allow teachers to step outside the stream of action and so, most often, the meaning-making takes place outside the classroom: in the hallway, in the staff room, on the journey home, over dinner or at teachers' conferences. On these occasions, teachers tell stories about their experiences, replaying events in a form which outsiders may dismiss as unreflective 'war stories'. Replaying the events of a school day and rehearsing alternative courses of action, however, is essential to making meaning of the experience.

In Johanna's case, there are many examples of such reflection. Indeed, one of the advantages for Johanna of our collaborative work was that she had someone to tell these stories to. Garth, her spouse, was very patient in listening to her stories but he didn't appreciate them in the way I did:

I have never had anyone who was really as interested in what I was doing as I was, and here was somebody who was totally interested! How many people does that happen to in life! I mean, can you imagine if someone came up to you and said, 'I really want to know all about you. Tell me in complete detail'. . . . No matter how much Garth tries, he could never be as interested in the actual machinations as you were. That was wonderful. It worked so well.

Such replays and rehearsals of experience may proceed with a technical, personal, problem solving or critical interest. When Shulman (1987, p. 19) talks about reflection, for example, he describes it as when a teacher 'looks back at the teaching and learning that has occurred, and reconstructs, re-enacts, and/or recaptures the events, the emotions, and the accomplishments' of teaching. It is through this process, Shulman points out, that professionals learn from experience. In Shulman's sense of reflection, the interest is essentially technical:

. . . it is likely that reflection is not merely a disposition (as in, 'she's such a reflective person!') or a set of strategies, but also the use of particular kinds of analytical knowledge brought to bear on one's work. Central to this process will be a review of the teaching in comparison to the ends that were sought. (Shulman, 1987, p. 19)

Other teacher educators and researchers, more concerned with a personal interest than with fidelity of means to ends, have encouraged teachers to tell stories about their work and lives in order that they may reshape their understanding of their past, present and future (see Carter, 1992; Clandinin & Connelly, 1991). Similarly, researchers with a critical interest, such as Tripp (1993) and Berlack and Berlack (1981) have argued that teachers should talk or write about their experience in order to understand it in new ways.

In our collaborative work, however, the larger part of the replay and rehearsal of classroom events was reflection with a problem solving interest. Johanna and I often talked in very concrete terms about the meaning of events we had just experienced, or about the possibilities for future action. These conversations often involved verbatim rehearsals of what we would say to a class. The vignette below describes a typical case of replay after the event.

Vignette 3: 'That's it for tricks'.

Picture us sitting in the quietest place we could find, the fire escape steps, interrupted from time to time by a class from the junior division of the school moving up or down from the playground. I had lately been reading Gramsci (1971), and was wondering whether his notion of common sense, that unreflective knowledge which is composed of both good sense and bad sense, might be a useful analytic construct for this study. Johanna had read a commentary on Gramsci and a working paper I had prepared on common sense, and thought it was all a little disconnected from reality. I had been talking about some of the common sense qualities of the teachers' knowledge I had noticed at the school and mentioned that I had been surprised how context-specific my own knowledge of teaching was:

Bill   It's like knowledge of what will work disappears if you change classes, or subjects, or — in my case — countries. Or if you have a bit of a break.

Johanna   Like summer? Which explains why I was terrified on the first day of school, although I have been doing this for twenty years. The break totally destroys your sense that you can do it again. The thing about teaching that every teacher knows is that it is like handling animals: the kids, if they know that you are scared, will be in control. So it's a confidence trick. It will work only if you can convince them that you can make it work. This is something that every teacher in their heart of hearts knows. Really, this is dicey at times. There is no way you could control a group of kids who didn't want to be controlled. It has to do with a tremendous number of tricks you pull out of your bag, things that you know about child psychology, and every teacher has their own bag of tricks. There is always a chance that those tricks may not work. You have to keep going through the bag.

My response to this image of teachers' knowledge as a bag of tricks was to think of several concrete cases where we had needed to dig deep into the bag.

Bill	With some kids, like Mark, you just get to the point where you think: 'My bag is empty. That's it for tricks'.
Johanna	Yes.
Bill	Then you get a case like Luke, just then. When you approach him, you don't know what's going to happen. You think that you can talk him into [rewriting his illustrated book], but you don't know whether he is going to pull a face and wipe you off for a month.
Johanna	Yes, because I had to watch his face really closely and judge what to say next. I was looking for the recognition in his eyes that he knew it was rubbish. That didn't come as quickly as I thought it would come with Luke. He obviously had more investment in the story than I thought.
Bill	I thought it would come as soon as I said, 'You are a smart guy, you can do better than this'.
Johanna	We managed to pull ourselves out of that, having taken the wrong tack to start with, to think of other tricks we knew.

* * *

In such ways, Johanna and I spent dozens of hours exploring our understanding of events we had been too busy enacting to reflect on at the time. Because our teaching was such a stream of unreflected experience, we needed to replay these stories in order to make meaning of the experiences we shared. These stories may not be very analytic, they were probably boring to the friends and spouses who had no experience of teaching, but they were stories which needed to be told if the experience were to contribute to our development as teachers.

Inquiry

Unlike *replay and rehearsal* and *introspection*, where reflection takes place at some distance from action, *inquiry* is a form of reflection which involves both action and discourse about action. More than this, it involves a process of deliberate movement between action and discourse. In the educational literature on reflection, the kind of deliberate process here called inquiry has been described and theorised by Carr and Kemmis (1986) in their work on action research. They describe action research as 'self-reflective spiral of cycles of planning, acting, observing and reflecting' (1986, p. 162). In addition to describing the form in which action research takes place, they link it with the critical interest for reflection and presume that the end in view of reflection is emancipatory educational and social change.

The form of reflection here called inquiry, however, may also be undertaken with technical, personal and problem solving interests. Grundy (1987, pp. 149–150), for example, describes a scenario where the action research cycle of planning, observation, action and reflection is followed, but the end in view is fidelity to already established practices of an activity based mathematics program. The descriptions of curriculum planning by Connelly and Clandinin (1988, pp. 170–186) also include a deliberate movement between discourse and action, but in the case they describe the inquiry is shaped by a personal interest. Inquiry shaped by a problem solving interest is well documented in Schön's work on reflection-in-action and in particular what he calls the 'reflective conversation', a process of conscious on-the-spot experimentation in the action-present which occurs when practitioners try to resolve the unfamiliar problems which confront them in their professional practice (Schön, 1987, pp. 26–31).

In the collaborative work Johanna and I did together in teaching writing and science, much of our reflection was shaped by problem solving interests. Unlike Johanna's music and art lessons, where she was working from a long-established repertoire, in science and writing we had to talk about each step before we could proceed. In science, especially, where neither of us had taught the subject before, we could not rely on what Schutz and Luckmann (1973) called 'the natural attitude', but had to find a way of connecting the new subject up with our established patterns of teaching. Inside this larger cycle of inquiry, we used all four of the forms of reflection which are identified in this paper. The way in which the other three forms may be embedded in a larger cycle of inquiry may be explored through a vignette drawn from our work together in science.

In this case we talked about the content of a lesson Johanna needed to teach and developed a lesson plan; Johanna taught the lesson; we talked about the detail of what she had done and connected this experience to an issue we had been talking about, Johanna's realisation that she lacked the content knowledge and commitment required to teach the sort of science recommended in the guideline.

Vignette 4: 'Science is a real mystery to me'.

This story begins one afternoon during a swimming lesson. While a specialist teacher took the swimming class, Johanna and I sat in the sun and talked about the next step to take in preparing students for the science assignment we had given them. First, we found the section in one of our text books that dealt with the assignment tasks we wanted to teach, classification of leaves and the life cycle of animals. As we talked about this material, Johanna wondered why it mattered that students would be able to classify things. We talked for a time about the arbitrariness of the act of classification, that the classification system chosen would then affect the way people saw the world, and the importance of having a system of classification in order to get the world's work done. I suggested that she do an activity like the one in the science curriculum guide, asking students to make up a system of classification to explain the objects in the room or a set of objects she chose for them. As we talked about this idea, we both became quite enthusiastic. We began rehearsing aloud the kinds of things we might say or do — such as having a student among the 'things' to be classified — and laughing together at the little jokes this might lead to. By the end, she said that this sounded like the sort of science lesson I'd like to do. It sounded like this to me, too, and I was sorry not to be there to do it.

I missed the lesson we had planned, but later I asked Johanna to tell me what had happened.

I didn't do any more planning for the class or thinking about it than our conversation at the pool. I had a vague idea of where I was going, what I was trying to do and it seemed to me like the kind of lesson I had done enough of that I could just fly with it. I realised that was wrong when I got into [the grade 7 class] and had to teach it. The instant that I began to teach it, I realised that it might be beneficial to pretend to be from another planet and to have collected these things we were trying to identify. So

I played that game with them, but it was the kind of thing that I really needed to have thought about ahead of time.

Although we had talked about what we needed to teach next and planned a way of presenting the ideas, when Johanna began to teach the lesson she realised that she had lost her grip on the lesson we had rehearsed. Instead, she followed an idea she had in class. The first time she used the 'man from Mars' idea it was with one of the relatively easy grade 7 classes and the game she played led to a reasonably successful lesson. When Johanna tried to repeat her success in the more challenging context of the grade 8 class, she realised that she did not really understand the idea of classification in science which she had set out to teach.

By the time I did it with the second class, the grade 8s, I realised that I wasn't really clear what I was trying to teach them about classification. I had a vague idea that what I was trying to do was to let them see that classification is arbitrary, depending on why you are doing the classifying. But that seemed to me to be a really hard concept to imagine the kids getting their heads around. I lacked the conviction that it was going to work. . . . I didn't know enough about classification. I couldn't give them examples of why someone would classify. What divides one species from another? I don't know enough science.

The lesson with the grade 8s did not go well, but the questions students asked helped her develop some of her own ideas about classification. By the time Johanna taught the third lesson, she had developed a clearer sense of the value of classification in science, and in the less demanding context of a small grade 7 class the lesson passed quite successfully.

Anyway, as I went through the three lessons I realised that there were some little things I could teach, such as that scientists actually *did* this, it didn't come from God, and that it had to do with ways of looking at objects and finding similarities and differences. Those things I could teach and I got better at teaching them.

We ended this small cycle of our larger inquiry with a moment of introspection. Moving beyond the replay of details of the lessons, Johanna talked about her lack of content knowledge and commitment in teaching science.

Johanna   It's so completely obvious that I don't know enough science and I haven't taught this way and done this kind of stuff to be a good science teacher.

Bill	So what's the difference between you and me in this. I obviously haven't taught science. Do I seem to have more depth of knowledge in science?
Johanna	I think you are more sure of yourself. You are fairly confident that you know as much science as you need to know to be able to help these kids. I don't. Science is a real mystery to me.

* * *

This story describes a brief cycle of inquiry, itself part of the larger reflective inquiry into teaching science. In this case, Johanna set out to teach a concept she did not really understand. By the time she had finished teaching three classes she had a stronger sense of what there might be worth teaching about the notion of classification in biology. She still, however, lacked the detailed content knowledge of examples from the classification systems biologists use and the purposes these systems serve. Within this cycle of inquiry there are examples of replay and rehearsal — in the lesson planning we did and in the stories Johanna told about her lessons — and of introspection. In addition, the process of learning through experience is also a form of reflection, called here, spontaneity, outlined in the next section.

Spontaneity

The fourth and final form of reflection distinguished here is *spontaneity*, the tacit reflection which takes place within the stream of experience. This is the form of reflection which corresponds with the 'jazz-player' form of Schön's reflection-in-action. In the midst of action, and without turning one's attention back against the stream of experience to become aware of this as action in the world of time and space, teachers seize the moment and change the direction of their action.

The tacit quality of spontaneity poses particular problems for a researcher hoping to document the phenomenon. In some cases, I noticed changes in direction which Johanna was only aware of after the fact. When we tried to talk about such cases more fully, the task of asking about awareness of a process which is by definition tacit led me to uncomfortable and inconclusive cross-examinations of Johanna. On one occasion, for example, I asked her how she came to make a major digression from the plan she had followed in the first of three similar lessons:

Johanna	In the second class I went into the telephone activity just because I was bored. I thought I'd basically cover the same things, but just do a bit of enrichment in the area of telephone interviewing.
Bill	So, did that occur to you in the break between the lessons?
Johanna	No, the moment came, I saw it and took it. It occurred to me immediately before you took it.
Bill	You actually did think, 'I will do it' before you did it?
Johanna	I probably did actually. I was thinking about Bob, actually, and a lesson I saw him teach last year on telephone interviewing. It occurred to me when I got to that section of the [Independent Field Trip] discussion that they were going to be doing a lot of phoning. I don't know when it happened, but I was doing it.

As an observer of the two lessons, it was easy for me to notice that there had been a change in plan. In place of a discussion about the rules for setting up an Independent Field Trip there was a role play of the telephone calls students would need to make. But when I asked how the change came about, Johanna was not at all certain what had happened. She might have thought about it, she could certainly remember seeing another teacher use the role play technique successfully, but she was finally unsure whether or not she had thought about it in advance. Because it was hard for Johanna to reconstruct an account of whether she did or did not move out of the stream of experience to reflect consciously on her options, the remaining examples in this section are drawn from my notes about my own teaching. Having wondered from the beginning of the study how to account for the role of thinking in the moment-by-moment inventions teachers' make in the classroom, I was more easily able to report on the phenomenon. In the vignette below, I describe a case where my spontaneous action led me to an unpredictable and unfamiliar place and I learned more about how to teach science to my grade 7 classes.

Vignette 5: 'Let's find out'.

During this lesson, students in the second grade 7 class were introduced to the notion that the formulation of problems is the first step in the scientific method. I used a series of activities from the school board handbook, *Science 7*, and began by asking students to formulate a problem based on their observations about the following:

1. Get down on your knees. Place an elbow against your knee and stretch your arm and fingers out on the floor.
2. Stand a chalk board eraser on the tip of your fingers.
3. Place both of your hands behind your back. Bend over and knock the eraser with your nose without falling over.
4. After doing the activity take a class vote for the results. How many males were successful? How many females?

I asked the class to read the instructions and write down whether they thought more boys or girls would be able to knock over the duster. Overwhelmingly, they seemed to think that it would be the girls. This was a surprise, and two possible explanations struck me. Either they had been talking about it in the hall to the other grade 7 class, who had just had the same lesson, or they had understood me to ask whether more boys than girls in this class would be successful. Something inside me, not quite conscious, told me not to check on the first possibility. If they had heard about it in the hallway, and they knew that I knew, it would seem silly to continue. Instead, I pointed out that there were more girls than boys in the class and asked whether people thought that girls would be more successful in general. This was still the prevailing opinion, but the fact that there were a few waverers made it possible to say, 'Well, let's find out'. So, following the same plan as the last class, I had one boy and then one girl try it until we ran out of boys. Once this activity started, the tone of the class completely changed. It was fun, there was lots of cheering, and from the beginning the activity took on a life of its own. Several girls could not do it, and several boys could, so I said that I was not sure that science was on my side today. In an aside to Johanna, who had been listening-in while she marked essays at her desk, I suggested that the effect of centre of gravity might depend on puberty.

The possibility that they could overturn the official expectations, combined with a boy-girl competition, made the activity lots of fun. After 12 students had tried, the pattern I expected had begun to emerge. I drew their attention to this, but Christine was unconvinced. She asked me to try, and I was able to knock it over. Next, Johanna tried and was not able to. One of the students picked up the puberty aside I had made, so plainly the effect of puberty on my 'centre of gravity' theory was in doubt. I asked people for a statement of the problem, and after several attempted explanations got, 'Why is it easier for women to lean forward than men?' I also asked for guesses about the cause of what we had seen, and got a range from the official centre of gravity explanations to the more frivolous suggestion that girls had longer necks and noses.

We then went on an activity involving vinegar, raisins and baking soda. When the baking soda was added to a conical flask containing a few raisins, the raisins were supposed to float to the top. I had planned to distribute the materials carefully to avoid confusion, but in the excitement of the duster activity I forgot my plan. Instead of dividing the class into groups and naming the person to collect each piece of equipment, I found myself surrounded by a dozen students clamouring for vinegar, raisins, water, conical flasks, graduated cylinders and measuring spoons. Johanna saw my confusion and put aside her marking to help me. At this point, I called the rest of the students in and began a demonstration around the bench on which I had stored the materials.

The first time, I swished the flask around too enthusiastically and the foam created by the reaction between the baking soda and the vinegar poured out over my arm and the desk. The students (and Johanna) seemed to like this, and someone said: 'This is what I call science!' Only one of the raisins rose up on the cloud of gas, so I tried again with less agitation, and two of the three rose. With the class gathered around the bench, I asked what the problem was ('Why does the raisin rise when the baking soda and vinegar were added?'), and explained that the reason was that the reaction between them created bubbles of gas which attached to the raisins and lifted them to the surface. I asked people why they thought some raisins rose and others didn't and was offered a series of suggestions relating to size, weight and surface area. It was now time to clean up, so the group gathered around the bench, broke up, washed the glassware and moved off to the morning meeting.

* * *

This vignette contains a series of practical teaching problems which were resolved spontaneously. When, for example, I was confronted by the class's surprising response to the duster activity — that they overwhelmingly thought that girls would be more successful — a series of possibilities flashed through my mind in a moment. Had they heard in the hallway from the other class? Did it matter if they had? What else could I do, anyway? I had no time to develop or consider alternative strategies, and I felt obliged by the pressure of my audience to move smoothly and confidently on to organising the activity I had announced, so I put aside the uncomfortable possibility that some of the students already knew what was going to happen.

The second example of spontaneity in this story concerns my decision to abandon the small-group focus of the activity involving raisins and vinegar. Having lost my mental place in the lesson plan, and briefly overwhelmed by the press of students around my bench, I spontaneously called the remainder of the class in for a demonstration of the experiment. Had I been teaching a more difficult class, this option would not have been open to me but with 19 enthusiastic and biddable students and the possibility of additional assistance from Johanna, I was able to seize the moment and develop a better lesson than I had planned. And in the process I learned that with a class as small, involved and tractable as this one, demonstrations can be at least as effective as small group experiments.

The third and final example of spontaneous reflection led to one of the best moments in my term of science teaching at Community School. The students liked the theatre of the conical flask overflowing down my arm, and they were curious about why some raisins would not rise. Rather than participating in a lesson where the students guessed at the name of the problem I had prepared, we framed a new problem from our observations and were left guessing at possible explanations. The genuine sense of inquiry raised by this unforeseen outcome was, I think, what the writers of the syllabus had in mind when they contributed their own tried and true lessons on problems, observations and inferences.

CONCLUSION

This chapter has outlined the range of reflection which emerged in an ethnographic study of teachers' learning. Reflection has been portrayed in terms of two dimensions: the interests and forms of reflection. The two dimensions were proposed as complementary, in that each of the interests may be served by each of the forms of reflection, and examples from the literature were provided for each of the possible combinations of interest and form. Some others who have written about reflection have connected a single interest with a single form of reflection, such as Carr and Kemmis' critical interest and inquiry. Other writers have connected a single interest with a range of forms: in Schön's case, various versions of his reflection-in-action and reflection-on-action were represented in all four of the forms of reflection, and are in each case associated with the problem solving interest identified in this framework. In the case study material, however, some interests and forms were more prominent than others: in short, introspection was most often with a personal interest; replay and rehearsal, inquiry and spontaneity were overwhelmingly pursued with a problem solving interest. There were a few examples where Johanna followed the critical interest through introspection, replay and rehearsal and spontaneity, and there were rather fewer examples which seemed to be informed by the technical interest in any form of reflection.

This framework of analysis emerged from the experience of the case study, and was not prepared until the empirical phase of the project had been completed. It was shaped by the intimate and introspective data available from a long-term, collaborative case study in which both the researcher and teacher were involved in reflecting on the experience of teaching. In addition, the framework was influenced by the preconceptions about teachers' knowledge and reflection I brought to the study, by the particular circumstances of the study, and by the relationship which Johanna and I developed. Because Johanna is such an open and forthcoming person, the introspective reflection tends to be personal. Because she is intuitive rather than rationalist, there was never any prospect that the study would document in detail the technical interest. Because Johanna sees herself as having reached a position of considerable freedom from external constraints in her teaching, she was unlikely to focus in great detail on the emancipatory interest of critical reflection. In a similar way, because I was more interested in how she and I solved the practical problems of the classroom than in reshaping the conditions of Johanna's work, it is not surprising that more attention was devoted to the problem solving interest than to critique. In other studies, teachers and researchers working in different conditions with different horizons of understanding about their lives and work might well engage in patterns of reflection which favour other interests or forms. These are possibilities which others may choose to explore.

The value of a typology such as the forms and interests of reflection developed in this paper is that it allows for a more subtle and textured account of teachers' reflection than either a Habermasian distinction among three interests or Schön's dichotomy alone can offer. The disadvantage of such typologies is that the effort to make the set of distinctions clear may lead to the impression that a particular typology is offered as the final word on the phenomena it describes. This ought not be the case with this conceptualisation of reflection. The categories in this typology are neither as separate nor as exhaustive as they may appear when represented as sixteen individual boxes on a chart. The categories on each dimension have been separated for the purpose of analysis; one interest of reflection may be dominant without the other interests being abandoned; and teachers may move from one form of reflection to another within a single conversation. By attempting to understand reflection as a participant in the action of teaching, however, it has been possible to offer a comprehensive account of the kinds of reflection teachers use as their understanding of teaching develops and changes.

NOTES

1. This chapter contains reworked material which first appeared in a 1991 book by William Louden titled *Understanding Teaching: Continuity and Change in Teachers' Knowledge* (London: Cassell & New York: Teachers College Press). It is reproduced here with the permission of the publishers.

SECTION IV

COLLABORATION

8. WORKING TOGETHER: AMANDA AND GEOFF

In earlier chapters, we have argued that teaching involves a search for a set of routines and patterns of action which resolve the problems posed by particular subjects and groups of children. Teachers' knowledge about these patterns, content and resolutions to familiar classroom problems are shaped by each teacher's biography and professional experience. A teacher's practice — shaped these historically sedimented predispositions to action — becomes a search for a more settled rather than a more effective practice (Wallace & Louden, 1992). Confronted by new problems and challenges, a teacher struggles to resolve them in a way that is consistent with the understanding that he or she brings to the problem at hand. In turn, this leads to a gradual reformation and growth of the teacher's knowledge.

In this chapter, we are concerned with the way in which teachers develop and apply their knowledge in the course of their work with colleagues.[1] While the history of educational policy and research seems to take a dim view of teachers' knowledge (Feiman-Nemser & Floden, 1986), it is clear that teachers' own stories help make their knowledge explicit, accessible and problematic. Critics of teachers' reliance on intuitive knowledge argue that this knowledge is narrowly held and that teachers' affiliations with one another rarely provoke close examination of their assumptions about practice. On the other hand, it is clear that knowing and talking about the problems of teaching is a powerful catalyst for improvement.

And yet, as Grumet (1978) reminds us, even telling a story to a friend is a risky business; the better the friend, the riskier the business. Friendship suggests a safe and confidential environment precipitating the telling of a deeper, more meaningful story than would otherwise be the case. But, sometimes the listener asks the wrong questions or makes different interpretations, appropriating only those parts of the story that he or she can use, ignoring the part that really matters to the teller. The story gets changed, shared with others who might be less sympathetic and played back in a form which is unacceptable to the teller. Once retold, the story adopts a life, a momentum and a meaning of its own, separate from the experience which generated it.

In the two narratives which follow, we explore how teachers are able to share their stories within the context of a collaborative relationship. The examples chosen are not held up to be 'ideal' but they were regarded as successful according to the judgement of the participants and the authors. The signals of success or otherwise of the collaborations were to be found in how well the participants managed the delicate balance between individual freedom, mutual respect and a desire to extend the horizons of their professional practice. Each story provides a general understanding (with some specific examples) of the relationship; its background, growth, philosophical and practical dimensions, and value to the individuals concerned. What worked[2] and what didn't in these relationships informs the later analysis of the qualities of successful collaboration.

AMANDA AND GEOFF

Amanda and Geoff were teachers at Brookvale Public School where Amanda taught a grade 1/2 class and Geoff a grade 2 class in a room across the hallway. Since joining the staff three years previously as young beginning teachers, they had developed a strong, friendly, collegial relationship. Others on the staff at Brookvale invariably referred to Amanda and Geoff in the same breath. They were regarded as having much in common: age, experience, grade level and teaching approach.

Philosophically, Amanda and Geoff were very close. Each favoured an experiential approach to teaching and learning (not uncommon in the junior grades) where students were encouraged to take some responsibility for their own learning and behaviour. They each translated these philosophies to match their own strengths as teachers. Amanda, for example, liked to use active learning centres for a number of subjects including science, social studies and mathematics. Her students were given a good deal of freedom to select and manage their learning within this framework. Geoff's teaching, by comparison, tended to be more teacher centred with Geoff providing the stimulus and direction for activities in science, music and mathematics through his talents as a teacher and classroom leader. Like Amanda, Geoff's teaching included a lot of encouragement for those students who were helpful, cooperative and who made a contribution to the well-being of the class.

Amanda and Geoff found numerous occasions to interact during the school week. Most days they combined classes for some joint activity. Geoff would lead the activities like singing because he played the guitar. Amanda organised the children and took other activities like story telling. Before and after school and at most breaks they would invariably visit each other's room for a chat about the happenings of the day and other professional and personal agendas. Generally, the conversation was the kind that would be expected between colleagues; friendly banter about children, classroom events, plans, school happenings, other colleagues, social life and extracurricular activities. Sometimes they also assumed the role of helpmate, making complimentary remarks and constructive comments about each other's program.

Amanda and Geoff seemed to value their relationship. Geoff summarised the benefits to him in the following way:

Amanda and I have worked very well together. I guess it was natural — we both came into the school at the same time. Last year she had a grade 1/2 and I took all of her 2s [this year], so more and more we are working together, planning programs together so that there has been mutual growth that way. Putting our two heads together saying, 'Let's try this or let's try that' and sharing things too — not just going off on your own. Saying, 'Hey did that work in your class — well it didn't work in my class'. What did she do differently than me? . . . We also have similar philosophies which is important because there are some people I simply couldn't work with because our philosophies on education clash so vigorously that there wouldn't be any opportunity to learn and grow.

These two teachers built a relationship based on mutual respect. In Geoff's words, their collaboration involved 'putting two heads together' to work, plan, try things, solve problems and to share hopes and dreams. Like Johanna and Bill, whose work is described in the next section, Amanda and Geoff's relationship served to make the work easier and altogether more enjoyable.[3]

JOHANNA AND BILL

Johanna was an experienced grade 7/8 teacher in an alternative public school and Bill — one of the authors of this book — a researcher with an interest in teachers' work and the process of teacher reflection. At the commencement of the study, Bill, himself an experienced secondary English teacher, approached Johanna (along with other teachers in the school) with a proposal for collaboration. Bill offered to make himself useful around the place and do some teaching for Johanna. In return, Johanna agreed to allow her practice to become the focus of Bill's study. Their association developed into a successful professional collaboration over a period of several months.

Johanna and Bill brought similar views about teaching and children to their relationship. Johanna's experience in teaching art and music in a progressive community school and Bill's past experience in teaching English meant that they shared a conviction that teaching was a craft rather than content oriented activity. They also shared views about the importance of striking the balance between providing students with the freedom to learn versus establishing high expectations for students to attain, a balance they called 'the education-control dilemma' (Louden, 1991).

Johanna used Bill as a kind of assistant teacher in the room. On occasions, Bill took over some of Johanna's teaching load while she worked elsewhere in the room. Although Bill spent a good deal of time in Johanna's classes and at times performed the same duties as Johanna, it was still clear to them both that Bill had different responsibilities and reasons for being there. As the relationship developed, Johanna and Bill began to share practical ideas about teaching, each learning from the other through their different subject backgrounds. Johanna explains how valuable she found it to be able to share her experiences with Bill.

I have never had anyone who was really as interested in what I was doing and here is someone who is totally interested. How many people does that happen to in life. I mean, can you imagine if someone came up to you and said, 'I really want to know all about you. Tell me in complete detail' . . . That was wonderful. It worked so well.

The trust that Johanna developed with Bill was quite personal in nature. They liked each other, became friends and the teaching became more than a piece of work for each of them. Bill enjoyed working with Johanna and the life of the school; Johanna liked having Bill around and hoped his research would go well.

QUALITIES OF COLLABORATION

Although each of these stories was located in a school setting, the collaborative relationships described here represent two very different situations (teacher-teacher and teacher-researcher). While much could be made of this distinction, this is not the focus of this chapter.[4] Of more interest here are those qualities which mark professional relationships which teachers find valuable and successful.[5] An examination of what worked and what didn't in the partnerships described will allow some of these qualities to be explored. Because success in human relationships is rarely an all or nothing affair, the list which follows should not be seen to be definitive or exhaustive. Combinations of qualities such as these will be present in successful relationships — for some purposes, for some of the time.

Similarities

Each of the partnerships described above was founded on similar traditions of teaching and hopes and dreams for education.[6] Attention was focused on essentially the same kinds of problems, around similarly aged children, dealing with similar subject matter, in similar settings. The importance of these similarities was that they allowed substantial intersubjective agreement about the meaning of classroom events. Amanda knew, for example, that Geoff's concerns about one of his science lessons were centred on the problem of student engagement because she shared his views on the importance of providing an enriching learning environment. Amanda had dropped into Geoff's class as he was teaching a lesson on the science topic 'Weather'. Immediately after the lesson it was clear that Geoff was worried by the fact that the students had not been actively engaged in the lesson. He raised his concerns with Amanda in the following way:

| Geoff: | I really wanted to get more done before Christmas but it is getting too busy to do things properly. It wasn't so good today. The kids didn't seem to be switched on. |
| Amanda: | With you it's a matter of knowing you rather than seeing what you are teaching. I know that you ordered the films and I saw that you had |

	ordered some books on weather and I saw that you have started the kids off on clouds.
Geoff:	There are a couple of things that I wanted to do that I didn't get done. Like the electro-board for recording the weather.
Amanda:	I agree that there is no point in gearing up to get started at this time of the year. You are better off getting started when you have time. And I thought you did an excellent job today.
Geoff:	I have gotten a lot of ideas from the David Suzuki book about how to integrate ideas about weather. I want to build up the vocabulary and do some more experiments with the kids. And maybe do some more work with air.
Amanda:	One idea that you may be able to use is to use a glass aquarium so that kids can see what is happening.
Geoff:	That's a good idea. Maybe I can set up a water table so the kids can explore for themselves.
Amanda:	Actually, I think I did something like that last year. I might still have the worksheet that I used.

Amanda was clearly tuned into Geoff's concern about student engagement in this exchange. She was able to provide some reassurance and support for his programming decisions, accept his self-criticism, offer some advice for him to build on, and finally provide some practical support to help him with his problem. Amanda's help, in this instance, represented a powerful and timely piece of collegial support. It was genuine, spontaneous, meaningful, focused on immediate problems and quite typical of the kind of talk-about-practice which characterised the relationship.

Like Amanda and Geoff's common understanding of the problem of student engagement, Bill and Johanna shared an understanding of the importance of the education-control dilemma. An example of this sharing is provided in the following lunch time conversation. Bill begins by asking Johanna about her writing program but the conversation soon turns to the dilemma in question.

| Johanna: | OK. I have learned a series of skills to do with helping students to write. I have added to my repertoire now. What we have done I would be able to do again, to refine it and get students to produce a story . . . First of all, to provide quiet in-class time. I have a whole variety of starting points from you. [Johanna goes on to list six of eight strategies that she had picked up from Bill.] I think I probably need to go over the notes for a lot more of the language because you've got things that you say to students that I need to memorise. |

Bill: Things that you hear and think, 'That worked?'

Johanna: Yes. With Group 3 I have learned the lesson of setting higher standards for them and letting them know that they are not meeting my expectations.

Bill: It came as a shock to me how simple it was to improve the quality of the stories. It was shocking that it was as simple as saying, 'It's not good enough'.

Johanna: Yes. [Laughs] That's the old argument in teaching. At what point does, 'It's not good enough' become discouraging? It's still a fine line but . . . It's much more devastating when it's said personally to one kid. We said it to a whole class. We said, 'Most of you are not good enough'. I think that helped us.

In this exchange, it was apparent that Bill and Johanna were tuned in to the same problem, that is, the education-control dilemma. Importantly, they also had a shared sense of how critical it was to tread what Johanna calls the 'fine line' between encouraging students and setting standards for them to meet. In trying to solve the problem they had each made a contribution, drawn from their own experience, to the other's learning.

In each of these examples, the participants were talking, not just about similar problems, but with a similar understanding of the importance of the problem and broad agreement about appropriate solutions. Without agreement at these fundamental levels, the participants could well have been 'talking a different game'. All teachers will have some similarities in their understanding of practice, but those alluded to here are much more subtle. It is these similarities which form the building blocks of effective communication.[7]

Differences

Differences create opportunities for partners to help one another across the gaps of understanding and to push the boundaries of practice. They also provide the spark for debate and reflection about practice. For Amanda and Geoff, differences were to be found in their distinct approaches to active learning. For example, Geoff particularly wanted to make his room a more stimulating place — something that Amanda had been able to achieve with her use of learning centres. His visits to Amanda's room provided much of the impetus and inspiration for his efforts. As Geoff reflected:

One of the things I was really looking to improve this year was the appearance of the room — to make sure that [the room] contains stuff [the students] can relate to, so I think that seeing Amanda's classroom helped a lot. . . . I'm not good at that sort of thing.

For Johanna and Bill, their different subject backgrounds often provided the catalyst for discussions about practice.

Bill: Our collaboration works so well because we agree on just about everything, but we teach different subjects . . .

Johanna: We have something to teach each other.

When Bill was sharing his ideas about language acquisition with Johanna, it was not an interaction between expert and novice or coach and student. Like Amanda and Geoff's conversations about learning environments, these were opportunities for collaborating professionals to pool their expertise to solve common problems and to extend individual understanding.

Symmetry

By symmetry, we mean balance or equality of power in a collaborative relationship. Successful relationships are based on fair exchange of workload and sharing of the risks involved in collaboration. All professional relationships, supervisor-teacher, outsider-teacher or teacher-teacher have the potential to fail on these counts because of competition between the parties. As Amanda points out, her efforts to collaborate with other teachers on the staff were less than successful.

When we started we thought it would be a good idea to change partners but when we asked people they [made excuses]. I really don't think that experienced teachers want to work with an inexperienced teacher. An experienced teacher thinks another experienced teacher has more to offer.

Each of the relationships described in this chapter was marked by the triumph of collaboration over the potential for competition. With Geoff, a colleague of similar experience and reputation, Amanda was able to build an equal relationship; both members of the partnership feeling secure in their roles as practitioner and helpmate. The contributions of Bill and Johanna to their relationship were different but equal. Although it transpired that Johanna sometimes adopted the attitude of the researcher and Bill did some teaching, there was no doubt that Johanna was the master teacher and Bill the researcher interested in her work. Through mutual respect, they were able to overcome the potential for rivalry in their relationship and retain their individual professional integrities.

Risk Sharing

A further quality of these successful collaborative relationships was a mutual preparedness to take risks and possibly fail in the presence of one's partner. Amanda and Geoff were constantly in and out of one another's classrooms exposing their practice to observation and possible critique. Johanna took the risk that the outcome of Bill's research was the careful documenting of her 'mistakes' and her relationships with other teachers. In her words: 'When we first started I had grave concerns about [having Bill observe] the school and my inability to work with Bob'. Bill took the risks of demonstrating his own inadequacy as a teacher in unfamiliar surroundings as he explained to Johanna: 'It's so funny! You have no idea what it is like recovering my repertoire. When I first started coming here I was scared to talk to kids as a teacher. I had forgotten how to do it'.

Without risk sharing, teachers are reluctant to expose their practice to others; instead they try to present a clinical image of 'coping well'. Teachers, with experience of the 'annual inspection', will readily recognise this kind of situation. Risk taking is important in teaching because it encourages experimentation with different ways of solving problems and, through reflection, increasing knowledge about practice. The presence of a colleague provides an audience for this reflection and increases the range of reflection. It was the luxury of having a colleague who was 'totally interested' in her work that Johanna found to be so valuable.

Trust

Like risk sharing, this quality embraces the notion that collaboration is a partnership in which there are no hidden agendas, and where teachers understand each other and respect each other's strengths and weaknesses. In the cases described here, the trust extended to genuine friendship of a

kind that is often found in the workplace.[8] In Johanna's words to Bill:
'You are my friend, and I am your friend. I would help you, we all do
that for each other'. And Amanda's words about Geoff: 'He knows that
if I say I am going to do something then I will do that . . . He knows that
because we are in and out of each other's classroom and helping each
other all the time'.

In these kinds of relationships, people look out for each other's
professional and personal interests and attend to the small matters of
courtesy and caring that characterise successful human relations. Trust
means accepting that, in Geoff's words 'putting two heads together' is
better than one and acting on conviction rather than constantly measuring
the value of the relationship to the individual.[9]

Emergence

Emergence allows for the relationship and its consequent benefits to
grow without being forced. Relationships which are most productive in
the long term are those which are unhurried and move back and forth
between and among professional learning agendas. Forcing the pace
through observation schedules, coaching cycles or preordained research
agendas often fails to attend to other qualities of collaboration such as
trust and risk sharing which take time to emerge. The relationship between
Amanda and Geoff, for example, had developed over a three year period
as a result of a natural growth in mutual interests and respect.

Geoff: I guess it was natural — we both came into the school at the same time
          . . . so more and more we are working together, planning programs
          together so that there has been mutual growth that way.

Their collaboration was not contrived nor forced by some external
program of peer intervention. Neither was support programmed by regular
schedules of observation and feedback. Rather, it was genuine,
spontaneous and tuned to the natural rhythms of the teaching day. Bill
and Johanna provided a different kind of case. What was important here
was not simply the time required for the teacher-researcher relationship
to develop (some months in fact) but Bill and Johanna's decision to allow
the research agenda to develop from their mutual interest in Johanna's
practice. This allowed for the other qualities of collaboration to emerge.

Humility

Teachers who demonstrate humility in relationships are able to find the balance between professional pride and professional modesty. In short, they have enough confidence in themselves and in the collegial situation to attempt things they do not know how to do. Without this quality, relationships can get stuck on the known and avoid pushing the boundaries of the unknown. If this had been the case, Johanna may well have been content to receive praise from Bill for her writing program, instead of seeking his constructive help. For the same reasons, Geoff may not have invited Amanda to observe and discuss his problems in engaging students in science. Humility, in Geoff's words, allows a teacher to 'try this' or 'try that' and reflect on those efforts in the company of a trusted colleague.

Fair Exchange

Fair exchange[10] means ensuring that the rewards of the relationship are balanced by the effort involved in maintaining it. Sometimes this can be achieved by simply sharing the workload on similar tasks, such as the preparation of worksheets. At other times, partners exchange particular expertise or qualities of value to the other person. Both Amanda and Geoff contributed particular professional and personal skills to make their relationship worthwhile for them both. In the case of Bill's relationship with Johanna, one reading might interpret it as unfair exchange, for Bill was clearly using the relationship to pursue his research agenda. This was not Johanna's reading however; what made the relationship fair for her was not only Bill's general 'helpfulness' but also his professional companionship and genuine interest in her problems of practice.

CONCLUSION

Emerging from this paper is a picture of the deeply personal nature of the qualities of successful collaboration in schools. Clearly, the qualities described here, such as trust, risk sharing, humility and fair exchange, can be found in any relationship between close friends. Why is it that professional relationships among teachers, in particular, are so personal? The way teachers teach depends on who they are. The meaning of their actions only becomes clear when it is set in the context of a teachers' narrative incorporating their personal and professional history and hopes and dreams for teaching. If teachers' stories have these very personal characteristics, then changes in teachers' knowledge are likely to take place in similar terms. Teacher collaboration, as an instrument of change, will only be successful when it attends to teachers' socially derived knowledge. As Christiansen and her colleagues (1997) point out, the process of collaboration is as important as the outcome.[11] Qualities such as emergence, trust and fair exchange are important because they are congruent with this notion; they allow time for the gradual reformation of knowledge through a recognition of the central place of biography and experience in teachers' working lives.

So what are the distinguishing features of a collaborative culture? The answer to this question does not lie in a simple checklist of eight behaviours. As a human activity, teaching is far more complex than this. Collaborative cultures have more to do with underlying principles of sharing, professionalism and democracy than checklists or action plans (Hargreaves, 1994; Wallace, 1998). Nonetheless, it is suggested that successful relationships exhibit certain attributes which we have called the qualities of collaboration. The eight qualities described in this paper were observed and enacted by the four teachers in this study. Whether all or some of these particular qualities extend to other collaborative relationships is not at issue; those judgements will be left to others. What is more important, we suggest, is the recognition that personal qualities, underscored by mutual trust and respect, form the basis for successful relationships in teaching and in research — operating as they do in different ways, for different purposes, for different people. As Connelly and Clandinin (1990) point out:

Collaborative research constitutes a relationship. In everyday life, the idea of friendship implies a sharing, an interpenetration of two or more persons' spheres of experience. Mere contact is acquaintanceship, not friendship. The same can be said for collaborative research which requires a close relationship akin to friendship. (p. 4)

For teachers and students alike, learning is a risky business, most likely to take place in a safe environment. As we have illustrated in previous chapters, for teachers, whose knowledge of teaching is closely connected to their biography, experience and sense of self worth, this is particularly the case. In the case studies described here, the teachers made a very real contribution to each other's learning. This process was genuine, spontaneous, proceeding slowly, by the gradual extending of horizons rather than by sudden leaps of insight. Collaboration in teaching offers a slow and powerful path towards educational change. It compels those who hope for change in schools to approach teachers with humility and respect.

NOTES

1. This chapter was jointly authored by John Wallace and William Louden. Some parts of the chapter are reproduced, with permission, from a 1994 article by the authors which appeared in *Qualitative Studies in Education, 7*(4), 323–334 (London: Taylor & Francis).

2. Reference is made here, and throughout the chapter, to the notion of a relationship which is 'working'. This term was used when it appeared that the relationship was serving the personal-professional needs of the individuals concerned and helping its members to examine and accomplish their work. 'Working' relationships were not simply harmonious, they also made progress by testing the boundaries of professional practice. The indicators of what worked and what didn't were to be found in the conversations and actions of the teachers. The judgements about these matters were made by us in collaboration with the teachers.

3. It is not the intention of this chapter to romanticise the idea of collaboration. We recognise the perils of contrived forms of collegiality (Hargreaves, 1994) and the need for teachers to work alone for much of the time.

4. We don't wish to underplay the differences between these two kinds of relationships. While we are aware of the different worlds (Bolster, 1983; Cuban, 1992) and unequal power relationships (Schroeder & Webb, 1997) in teacher-researcher relationships, we argue that similar issues arise in teacher-teacher relationships (particularly between administrators and teachers).

5. What distinguishes this list of qualities of collaboration from others is that it is drawn from empirical evidence. If these qualities appear to be a little more than common sense, it is not surprising since all relationships derive from human experience.

6. McWhorter and her colleagues (1998) describe how agreement on basic philosophies of teaching provided a 'touchstone' for decisions made within a collaborative team.

7.   McIntyre's (1981) notion of narrative unity suggests that people's relationships are founded on common, similar or shared experiences.

8.   These relationships could be interpreted as more than professional. This is a reasonable observation and not entirely untrue. Relationships are not generally partitioned as entirely either 'professional' or 'personal'. These relationships are primarily professional, but they are built on personal foundations.

9.   A relationship of trust is analogous to a culture of care (Noddings, 1984) where 'we must commit ourselves to the openness that permits us to receive the other' (p. 104). Trust is a central theme in much of the writing about relationships in teaching, for example, Christiansen *et al.* (1997) and Schulz (1997).

10. Fair exchange, along with symmetry and balance of power are components of what Schulz (1997) called 'genuine mutuality'.

11. Nias and her associates (1989) go further, asserting that the preconditions for success in any program are to be found, not in the qualities of the program, but in the qualities of the collaboration.

9. ETHICS AND RESEARCH: MALCOLM

Whereas the previous chapter draws directly on empirical data from our research into collaboration in schools, this chapter is situated within the context of our broader research experience and our reflections on that experience. Over the years, our understanding of ethical considerations in research has evolved alongside our use of different forms of inquiry. As doctoral students, we learned much about the potential of narrative as a means of capturing the lived experience of those involved in schools. However, when we began to incorporate these methods into our own research, we found ourselves transferring the old canons of quantitative research to the new genre of qualitative research. Issues such as access, informed consent, security of data and anonymity of participants seemed to be the dominant considerations. With experience, we began to understand that narrative represents the particular reality of the storyteller, a reality constructed for the purpose of telling the story. This kind of research raised ethical issues of a different order involving the fundamental relationship between the researcher and the researched.

For the past several years we have worked in university centres for teaching and research in education. We now find ourselves using narrative methods in several strands of inquiry, each with its own particular ethical consideration. Firstly, we use narrative as a technique for our personal research agenda, as a way of presenting the voices of teachers and students to support our arguments about schooling. Secondly, we use narrative in funded program evaluation research as a way of reporting the progress of a particular innovation. Thirdly, we use teacher-written stories and commentaries on stories to elaborate the dilemmas of teaching and to help teachers reflect on their practice. While many ethical issues are common to each of these kinds of inquiry, others are specific to particular forms and often particular cases. In the first form of inquiry, questions of ownership and voice are often paramount. The second form of inquiry is often dominated by a consideration of the agenda of the funding body.

In the third form of inquiry, questions of anonymity vs visibility of participants are important.

In the broadest sense, ethics is concerned with the treatment of individuals in the research process. At first glance, the qualities of collaboration discussed in the previous chapter — trust, symmetry, risk sharing, humility — are the important values undergirding the relationship and the research activity. However, research is also supposed to advance the knowledge base about teaching, learning and schooling. With the advent of narrative forms of inquiry in recent years, two issues have become problematic. The first issue concerns the nature of knowledge, understanding and truth, and second involves the rights of human subjects in the research process.[1] In the analysis which follows, we explore how these two issues are played out in practice and how they relate to one another.

Perhaps the most important message from our work in narrative research is that ethical issues are complex, as complex as the nature of human relationships themselves. Each case is different, calling for different solutions depending on the personal convictions of the participants. These issues, according to Lincoln (1990), are subtle and particular — about 'particular persons in particular contexts at particular times, with many persons acting on different sets of personal and professional principles, depending on who they are and where they sit in the context' (p. 285/6). While these behind-the-scenes episodes are often untold, they are an important part of the story of doing research. With this aim in mind, we share four research episodes requiring ethical resolution — three are from our own narrative inquiry and one is taken from the experience of a teacher colleague 'participant' in a research study. The first episode, which took place during John's doctoral research, concerns how one teacher, Malcolm, took exception to John's depiction of him as a conservative teacher. The second episode involves Simon, a teacher who was left feeling betrayed after his involvement in a research project. In the third episode we tell the story of our involvement in an evaluation of the implementation of a new physics syllabus. The final episode examines some of the ethical issues encountered when we recently edited a casebook of teacher stories. Through an examination of these episodes, we assemble some case knowledge to guide narrative research.

EPISODE 1: MALCOLM

Malcolm was one of five teachers who agreed to participate in John's doctoral research study into the effect of peer supervision on teacher growth in a Canadian elementary school. The story, and the subsequent episode about Simon, is told by John in the first person.

I was initially invited to the school by the school principal who was a basketball teammate. The principal introduced me to five teachers who agreed to participate in my study. In separate interviews with each teacher I explained the kind of activity that I was interested in observing and the extent of commitment required of them. In the ensuing weeks, I observed the classes of each teacher, conducted interviews and, where practical, sat in on the formal and informal discussions between the study teachers and their nominated peers. I began the study with the intention of spending approximately one day each week with each of the study teachers. However, as the study proceeded, I found myself spending more time with some of the study teachers than with others. Although I did not admit it at the time, I had developed in my mind a hierarchy of preferred teachers — two young 'progressives' at one end of the scale and two older 'conservatives' at the other end of the scale. My views were undoubtedly influenced by my close relationship with the principal who often expressed his concerns about those members of staff who were more 'traditional' in their teaching approach. However, I was still confident that my superior research skills would overcome any biases I might have about the teaching abilities of any of the study teachers.

It was my habit at the end of each day to record a narrative account of my impressions in the form of field notes. About five weeks into the study, I decided that it was time to share my impressions with each of the study teachers to gauge their reactions and obtain feedback. I carefully separated my notes into five separate accounts, one pertaining to each of the five teachers and asked each teacher to read my account and provide me with feedback on what I had written. My notes about Malcolm contained several paragraphs of interpretive comment including:

The interesting thing about this interview for me was the fact that Malcolm kept returning to the theme of the dilemma he is facing regarding his individuality as a teacher and how he sees that being threatened by the pressure to conform to new trends (such as active learning). This is especially difficult for him because he has been teaching in the school for some years during which time his personal teaching philosophy was the accepted way of doing things. Now it would seem that this is no longer the case — he finds himself being questioned (by the Board philosophy, some colleagues, and the administration) in subtle ways and as a result has begun to question himself. His professional growth is inextricably tied to this dilemma. His strategy is a personal one — to explore ways of making his classroom as interesting as possible but retaining the basic philosophy of a framework of structure to teach the fundamentals of reading and math. However because of the pressures around him he finds it prudent to do this in isolation because of the quiet fear that to open his classroom up to observation may invite unwelcome criticism from his colleagues.

The following day I arrived at the school to find a package from Malcolm in my mail box. The package contained several pages of a closely typed response to my notes. He began his response with the following:

Although I agree with much of what you have written, I disagree with many of the assumptions you have made. As teachers, we are all too ready to play amateur psychologist, when we are unaware of the factual background. As a person interested in science, I would think that you should be particularly careful to check the facts before coming to conclusions.

Malcolm went on to apprise me of the 'facts'. In a detailed account of his 25 years as a teacher, he described his experience with active learning, his history of leadership in various settings, his work with colleagues, his concern for students, his progressive philosophy and his standing in the school community. He also included copies of annual teacher evaluations forms going back over several years attesting to the excellence of his teaching. Malcolm concluded his reaction as follows:

I feel that you have attacked my integrity with your comments and so I have had little choice but to defend myself by presenting the facts. Naturally I resent your suggestion that my reluctance to put myself forward for a position of leadership may stem from a fear of pressure, a reluctance to conform to new trends, a desire for isolation or a fear of unwelcome criticism from colleagues. My reply to your remarks will, no doubt, be considered defensiveness or conceit on my part. You will be free to have the last word, interpreting my remarks and editing them. However, I feel far more confident of your ability to make objective interpretations once you have some background material.

After reading Malcolm's response, I was fearful that I may have done irreparable damage to our relationship. In my field notes I had tried to be as honest as possible by sharing my interpretation of the events that I had observed. It appeared that Malcolm not only disagreed with my interpretation but took it as a criticism of his teaching abilities. Later I went to see Malcolm about his response, worried that he may wish to withdraw from this study. It was, however, a cordial meeting. Malcolm reiterated his concern that I needed to get the facts straight before jumping to conclusions and I assured him that I would read his notes over again and discuss them in detail with him later. I was pleased that he seemed happy for us to continue to work together.

The study proceeded with Malcolm as a participant although this incident had a sobering effect on our continuing relationship and on the quality of the data collected. I spent less time in Malcolm's classroom than before and restricted my data gathering to formal interviews and observations of prearranged conferences between Malcolm and his partner. My subsequent field notes about Malcolm consisted mainly of transcripts and descriptive material rather than interpretive analysis. In my dissertation my portrayal of Malcolm is still closer to my original view of Malcolm as a 'basically conservative teacher' than to Malcolm's own public view of himself (although I acknowledged these differences by footnoting Malcolm's objections to my position).

The memory of this experience from early in my career as a researcher is still fresh in my mind. It is clear that my portrait of Malcolm did not match his own perception of himself; he was less than pleased with the images I reflected back to him. At the time, I thought that my image of Malcolm's practice was important for the story and it seemed to be well grounded in the data. I had no special malice towards Malcolm; I thought that I was doing my best to see the world from his perspective. Yet, from his perspective, I came up short. My narrative of Malcolm was not the same as the narrative Malcolm would tell about himself. While much has been written about the importance of shared meaning in narrative research, it is inevitable that in some circumstances, meanings will not be shared at all. So, whose story should take precedence in these situations? What obligation does the researcher have to capture the voice of the teacher? Does the researcher have an equal obligation to produce an alternative account? What does informed consent mean in such situations?

EPISODE 2: SIMON

Some time after completing my doctoral studies I had the opportunity to view some of the issues I had encountered with Malcolm from another perspective. I ran into Simon, a colleague from my high school teaching days, at a sporting event. We began talking about old times and the progress of our respective careers. Simon spoke about his current teaching situation and I told him about my latest research into science classrooms. This mention of research prompted a strange reaction from Simon and he began to tell me of his experience as a teacher participant in a recent research project.

Simon had been approached by a researcher with a request to conduct research in his high school science class. The study was conducted over several weeks involving intensive classroom observation and interviews with both Simon and his students. During the interviews, Simon recalled that he was asked questions about many aspects of his teaching but he particularly remembered being quizzed about his relationship with the girls in the class. At the time he did not place great significance on this aspect of the study because the researcher revealed little about his own interpretation of the meaning of events. Some time after the study had finished, Simon was shown a copy of a published account of the study by a friend. He was incensed to find that the researcher had written a largely unflattering account, highlighting the sex-biased aspects of his science teaching practice. Although this experience occurred several months before our encounter, Simon was clearly still scarred by the experience. In his words to me at the time:

I feel a degree of betrayal. At no time was the research agenda made perfectly clear to me. [The researcher] seemed to gloss over what he was about. I mistakenly interpreted this as meaning that if we (I) had a good idea what he was about that this would somehow bias the data that he was collecting. I always believed that a good teacher has nothing to hide — that everyone should be welcome in his/her classroom. Unfortunately, I no longer believe this. In investigating gender differences (as in hindsight I now understand) the researcher queried what was happening with respect to my classroom and questioned decisions that I was making with respect to them. In the published account I resent the interpretation that was placed on aspects of my behaviour as a classroom teacher — in particular, the continual reference to sexual bias in my teaching. In my fifteen years of teaching it has never been suggested that I was a party to the kind of conduct suggested by this researcher. In hindsight I believe that I was a little frank and embraced his desire to know my thoughts on teaching as it now appears to me that which supported his notions was seized upon and that which did not fit his preconceived notions was ignored.

This account by Simon of his experience on the 'receiving end' of a narrative research study provides a different view of the ethical dimensions of research. One is immediately inclined to be sympathetic with Simon's predicament. In a gesture of goodwill, a teacher with a long history of success in the classroom invites a researcher into his classroom. The teacher, flattered by the attention shown to him by the researcher, is open and frank about his methods. He reveals the strengths and weaknesses of his teaching and shares his thoughts about the reasons for his actions. However, at the end of the study, he feels betrayed by the researcher's depiction of him as a sexist teacher. The story told by the researcher is not the story the teacher would have told about himself.

Is this a clear case of unethical behaviour on the part of the researcher? Perhaps, but perhaps not. While we don't have the researcher's account of this incident, it is probable that he did in fact obtain Simon's consent to proceed with the study and, as far as he was aware of his intentions at the time, he may well have informed Simon of his research focus. However, researchers are not always sure of their focus and aware of their preconceptions. It is possible that the issue of girls and science may not have emerged until later in the study. Even if the researcher had shared his field notes with Simon, the gender issue may have been hidden amongst reams of other descriptive data. Maybe the researcher left the site with several possible avenues for the analysis of this data.

What are the researcher's responsibilities beyond this point? He could have chosen to consult further with Simon to arrive at a 'shared' meaning of the events described. However, shared meanings are only possible when there is close philosophical agreement between individuals. In this case, it is unlikely that Simon would have agreed to an account highlighting his possibly sexist teaching practices. However, such accounts are important if we wish to make progress on the problem of the gendered nature of science classrooms. In proceeding with publication the researcher chose a particular course of action. He probably reasoned that he had fulfilled the responsibilities of informed consent through his work with Simon during the data collection phase. In his published account, he constructed a narrative describing the sexist practices of an anonymous teacher based on raw data which he had collected in the classroom of a teacher code-named Simon.

EPISODE 3: MR WARD

This episode examines some of the ethical considerations surrounding our research involving Mr Ward, discussed previously in Chapter 4. A few years ago I was involved with our colleague Helen Wildy in an evaluation of the implementation of a new state-wide physics syllabus with constructivist underpinnings (Wildy & Wallace, 1995). The new syllabus replaced a highly theoretical course with a context-based approach where physics ideas were introduced through familiar practical situations and experimental work. Our evaluation involved a comprehensive survey of several hundred physics teachers from throughout the state as well as several narrative case studies of physics teachers attempting to incorporate the new teaching and assessment strategies. Narrative accounts were constructed by observing each teacher's classroom over several lessons and interviewing the teachers and their students. We found teachers in various stages of implementing the change — some were beginning to tinker with the new strategies while others had made substantial changes to their practice. In almost every case, we were able to report teachers' approaches to the new syllabus in a positive light. The one exception to this generally positive outlook was Mr Ward.

Mr Ward had an honours degree in physical chemistry. He was head of science at an elite private school and had taught physics at high school level for more than 20 years. Previously he had been a university tutor and lecturer for several years. He was well known and highly respected for his teaching, both in his own school and within the community of science teachers in the state.

We studied Mr Ward in his physics class for several weeks. We observed that his teaching was very much like the type of science teaching that the new syllabus was designed to overcome: an emphasis on coverage over understanding, teaching to the examination, whole class instruction, failure to take account of students' prior knowledge, development of algorithmic knowledge and skills, isolated theory and unfamiliar contexts. Our initial reaction was one of dismay. Mr Ward appeared to be making little concession to the tenets of the new syllabus. Further, he seemed unapologetic about his rejection of its philosophical underpinnings. We were therefore faced with a dilemma: how to capture Mr Ward's narrative in a manner which was faithful to his position without appearing negative or unduly patronising.

As we considered this dilemma and talked more with Mr Ward about his beliefs about teaching physics, it dawned on us that Mr Ward was adopting a position with respect to the new syllabus which, though unfashionable, had its own particular legitimacy. We began to understand Mr Ward's arguments for teaching the way he did. He was an experienced physics teacher in a school from which large numbers of students were selected to study at university level. His own physics content knowledge was extensive in breadth and depth. He was confident in a set of teaching strategies that worked for him. Confronted with a major syllabus change, he initially tried a few of the constructivist teaching strategies. However, he soon rejected the strategies as inappropriate because they did not produce the evidence of learning he had anticipated. He did not feel comfortable, neither did some of his students: they failed to learn the basic physics concepts and lost confidence. Some students even left the course. To Mr Ward this was enough to convince him to revert to his former well practised strategies. It seemed that making connections between physics and the real world was not as important — for these students, their parents and the school — as gaining high examination scores.

In Mr Ward's narrative, we were forced to change the direction of our gaze. Whereas previously we looked at teachers' practices through the lens of the new syllabus and characterised teachers accordingly, in Mr Ward's case we looked at the new syllabus through the eyes of Mr Ward. The result we believe was a much richer — and ethically more defensible — account of Mr Ward's practice than might have been otherwise achieved. By adopting the 'natural attitude' of the teacher in this case, we produced an account which became a centrepiece of the evaluation and take-off point for discussion about the efficacy of the new syllabus.

EPISODE 4: TEACHERS' STORIES

The final episode concerns some of the ethical issues encountered when we edited a casebook of teachers' stories (Louden & Wallace, 1996).

The idea for the casebook came about during some consultancy work on teacher competency. We were collecting teacher-written cases to exemplify components of the Australian National Competency Framework for Beginning Teaching (Louden & Wallace, 1993). As we undertook this work we came to appreciate the power of stories in teacher professional development. The development of a casebook of teacher stories employing a similar model to that pioneered by Judith Shulman (1992) seemed like the next logical step. In our work with graduate students, we had already assembled a core of good stories and invited commentaries. We supplemented these stories and commentaries with others written by teachers during a vacation institute on narrative methodologies. In total, we assembled 23 teacher-written stories with each story accompanied by two or more commentaries. The stories covered a range of accounts of successful and less successful lessons, interactions with colleagues and parents, and the rewards and frustrations of teaching.

As we set about the task of assembling the casebook, we found ourselves moving through several layers of ethical issues. The first issue concerns the anonymity or visibility of the authors of the stories. In the first draft of the casebook, we decided to list author names as a group on the title page and leave individual cases anonymous. The persons and schools mentioned in each case were given pseudonyms. However, this technique of preserving the anonymity of informants — standard practice in most narrative accounts — did not seem to be appropriate in this circumstance. We had the additional problem of how to treat the names of the commentators — in most cases close colleagues of the teachers who had written the stories. The transparency of this method was soon revealed to me when I used an 'anonymous' story during a teacher professional development session and one of the participants announced that this story took place in her school.

After consulting with some of the teachers, we abandoned this approach and decided to include the full author's name against each case. However, while this method seemed ethically more honest, it raised additional issues for the teacher-authors. One of the authors had written about her experiences as a homework tutor to her nephew. To collect data for the story, she had tape recorded a homework session. However, after transcribing the tape and writing the story, she was personally dismayed at some of her teaching practices and decided that she was too embarrassed to reveal her inadequacies, particularly to family members. She decided that her name should be removed from the story and that it should be edited to remove any possible association with her family. Other teachers had also written about potentially contentious situations and, although willing to be attributed authorship, were careful that their text excluded reference to specific persons or schools.

A further issue concerns the ownership of the cases. As editors of this volume, we exercised our editorial discretion over the stories. In most cases, we believed that the readability of the stories was enhanced by our copy editing. In a few cases, where meanings were unclear, we also put our own emphasis on the stories. In every case we sought the written permission of the original authors before proceeding. Most authors were prepared to sign off their permission without consulting the other persons involved. Some of our authors took this responsibility most seriously and followed a rigorous consultation procedure to ensure that other members of the school community (including the commentators on the stories) were happy with the way they had been represented.

In only one case did a school withhold permission for a story to be used. The site of the story was an alternative community school with history of collective and negotiated decision making. While the staff had initially consented to a story being written about the school, they had several concerns when they sighted the story itself. It was clear that the process of obtaining final permission to publish the story would have involved the author (and the editors) in a period of prolonged negotiations. Bill and I were reluctant to drop this story from the volume because we believed that it provided an excellent (and balanced) treatment of the issues involved in teaching in such an environment. However, with the publisher's deadline fast approaching, we felt that we had neither the time to negotiate changes nor the moral authority to publish the story. The story did not appear in the casebook.

We began the casebook project imagining that the standard ethical procedures of anonymity and informed consent could be applied in a blanket way. We soon learned that each case required individual consideration. The 'ethical complexity' of each case was influenced by a myriad of factors including the relationship between the editors and the teacher-authors, the relationship between authors and colleagues, the subject matter of the case, and the professional experience and confidence of the authors.

CONCLUSION

In the foregoing, we have re-created some situations that we found troublesome during our careers as narrative researchers. They were real, unexpected, and in some cases unresolvable, and provide some feel for the kinds of issues that can be encountered in the complex arena of research into human affairs. Embedded in these experiences are a string of ethical questions. From the cases of Malcolm and Simon, one might ask what happens when the researcher's portraits do not match people's own perceptions of themselves? What does informed mean and what does consent mean within the principle of informed consent? When does informed consent start and when does it finish? Is it always possible or desirable to seek a shared meaning in a narrative account of a teacher's work? Simon's case leads us to consider the difference between 'factual' and 'fictive' accounts and how ethical responsibilities to participants might change when alternative genres are used in research texts. Do researchers have a public and professional duty to describe situations of perceived social injustice? Both the Simon and Mr Ward incidents raise questions about how contentious issues should be depicted. The teacher casebook raises questions about the anonymity of participants and the rights of the secondary participants (students/colleagues) in narrative accounts.

The temptation here is to deal with each one of these issues in turn by proposing a set of solutions which might be applied across all cases where such problems are encountered. We will resist this temptation. Rather, we extract some of the broader lessons which emerge from our experience.

The first lesson concerns the fundamental nature of relationships between individuals in research of this kind. In each of the incidents described above, the ethical dilemmas and their resolution were a product of the people concerned. The argument here is that ethical decisions should be guided by a concern for nurturing and maintaining human relationships. In the previous chapter we argued that caring relationships in research include such qualities as similarities, differences, symmetry, risk sharing, trust, emergence, humility and fair exchange. In this approach to ethics — which Noddings (1988) calls relational ethics — the context of the situation and the individuals involved are no longer considered to be messy variables that hinder the uniform application of ethical principles (Brickhouse, 1992). A deep understanding of individuals and their circumstances thus becomes the way forward for resolving ethical dilemmas and, simultaneously, developing contextually rich narrative accounts.[2]

The second lesson concerns the utility and limitations of technical ethical procedures such as informed consent and confidentiality. There is little doubt technical criteria provide a more than useful starting point for narrative researchers. It is, for example, entirely reasonable to expect researchers to be scrupulous in their dealings with teachers and open about their intentions for the research. However, as the research process moves from the field itself to field texts and then to research texts (Clandinin & Connelly, 1994), contact with the participants diminishes and the researcher becomes less concerned with accurate description and more concerned with theorising. As Simon's experience illustrates, informed consent becomes practically and intellectually more difficult under these circumstances. Hence, the researcher must look for other guiding ethical principles to protect the interests of the participants. Similarly, the technique of confidentiality is well founded in the maxim of individual privacy and protection. However, as we discovered in the casebook project, privacy and anonymity are not always possible in narrative case studies of high fidelity (Lincoln & Guba, 1989). Even in the city of one million people in which we live, it is a relatively straightforward task to track down the source of research data. Perhaps researchers can deliver deniability but not anonymity. Moreover, as we strive to capture the lived experiences of individuals in schools, the participants are likely to have high ownership of the stories told and of the research experience itself. As our experience with the casebook illustrates, anonymity is not always desirable or attainable. Also, people may initially provide formal consent without realising the full implications of this process.[3] Roles are not static and require continuous negotiation and renegotiation of the relationship (Schroeder & Webb, 1997). Hence, we argue that technical ethical procedures need to be enacted within a set of higher ethical principles based on relationships between individuals.[4]

The third lesson concerns the issue of how to represent those teachers whose perspectives may threaten or challenge our own. Hargreaves (1996) has argued that narrative researchers naturally drift towards articulate, literary, humanistic, child-centred teachers whose values and style reflect those held by the researchers. When researchers encounter teachers whose values and orientations seem alien, these teachers are often depicted as being out of step, traditional, difficult and resistant. Studying teachers such as Malcolm, Simon and Mr Ward provides a challenge to make meaning of their actions and perspectives in terms of the circumstances they encounter and the lives they lead. And yet, this approach also holds some dangers for narrative researchers. The closer that researchers get to their subjects, the more difficult it becomes to describe unacceptable behaviour in those terms. It is relatively easy to hold on to firm judgements about people from a distance, but more difficult when you engage closely with them. Hence, conclusions reached by narrative researchers can often be seen as an artefact of the method used rather than the product of unbiased judgement.

A fourth lesson concerns the issue of voice in narrative research. As Clandinin and Connelly (1994) point out, both researcher and participant are confronted by issues of voice. When texts are created by one person about others, the writing process brings some voices to the fore and hides others. The researcher in Simon's case, for example, faced an ethical dilemma in attempting to speak to his audience in his own voice while trying to present an authentic account of Simon's voice. In Simon's view, the researcher failed on the second count. In the Mr Ward case, on the other hand, the final published account was criticised by some readers as being too soft on the teacher. In these situations, researchers are operating in what Schön (1987) refers to as 'indeterminate zones of practice' characterised by uncertainty, uniqueness and value conflict. They are, in a sense, caught in the middle between participant and audience. These circumstances challenge the most experienced of researchers.[5] Such dilemmas cannot simply be resolved by applying theories or techniques but rather by using the researcher's professional judgement based on past experience of similar problematic situations and personal ethical standards.

This issue of voice is further complicated by the use of fictive accounts where the voices of several participants are merged to form a picture of a composite teacher working in a composite classroom. This technique was employed by one of the writers in the teacher casebook project who wrote a story about a fictive social studies teacher based on her collective experience working in many schools. Under these circumstances, it is difficult, if not impossible, to trace the voice of the teacher to its source or sources. The voice in these kinds of texts becomes more the voice of the writer or researcher as he/she weaves the story together to form a plausible and interesting account. This technique of fictive writing creates a distance between the field and the various texts. This may help to release the researcher from an obligation to consult with participants or it may necessitate complex negotiations with each participant before the text can be finalised. Again, the resolution of these ethical questions is a matter for the researcher's professional judgement on a case-by-case basis.

This foregoing set of lessons do not exhaust the complex layers of issues encountered when conducting narrative research. In this book, we have argued that teachers and researchers can learn much by examining the rich material that emerges from the narrative process. However, it is fraught with difficulties. Most of these difficulties stem from the relational nature of the research process. The relationship between participants, researchers and audience influences the stories told, the interpretations made, and the final published texts in fundamental ways (Clandinin & Connelly, 1994). The relationship determines the nature and degree of collaboration and consultation around the formation of the various texts. Equally, narrative research is fraught with the kinds of difficulties which often beset human relationships — such as misunderstanding, exploitation, betrayal, insecurity and competition. As narrative researchers we work in an uncertain domain where there is no consensus or unanimity on what should be private and public and what might constitute harm (Punch, 1994). Nevertheless, it is our responsibility to conduct our research within an ethic of care. And even when we are careful, it is easy to get it wrong. Methodology, ethics and rigour are therefore intertwined in narrative research. It follows that research which is rich and interesting requires complex and careful ethical treatment. In finding a way forward, it is important to recognise the limitations of rules and maxims and to focus instead on the particularities of individual teachers and their circumstances.

NOTES

1. Some academics situate these two issues — about textual authority and about interpersonal transactions — as part of the postmodern debate about power and authority (Britzman, 1991; Brodkey, 1987; Lather, 1991).
2. This argument is supported by several recent elaborated accounts of teacher-researcher relationships in Christiansen *et al.* (1997), Clandinin, Davies, Hogan and Kennard (1993), Cochran-Smith and Lytle (1993), Hollingsworth (1992), Miller (1990) and Schulz (1997).
3. As McAlpine and Crago (1997) point out, this problem is accentuated when research is conducted across cultures.
4. For a more detailed discussion of the ethical tensions and dilemmas in these conventional ethical procedures, see Schulz (1997). Also Clandinin and Connelly (1988) provide an example of how ethical procedures sit within the context of ethical relationships between teachers and researchers.
5. Smith (1990), for example, writing of his involvement in a collaborative research project, describes how he 'did his best' but still 'came up short' in the eyes of the participants (p. 266). Miller (1992) talks of ongoing issues of power and authority among participants in the years following the publication of her book (Miller, 1990).

SECTION V

CONCLUSION

10. TEACHERS' LEARNING AND THE POSSIBILITY OF CHANGE

This book has brought together stories we have been collecting for almost a decade. The stories are all about science teaching, but they have more in common than the teaching context. They have in common also a view about the nature of teachers' learning; how teachers know what they know, how their knowledge changes, and how their knowledge is influenced by changes in subject matter and milieu. In addition, the stories have in common a view about the value of reflection, collaboration and ethical relationships between teachers and researchers. Throughout the book, we have circled around and often returned to these key themes. The purpose of this chapter is to bring together and summarise our arguments on each of the key themes, in order to make some claims about the possibilities for change in science education.

For researchers working with narrative case studies of individual teachers, there are risks in building a more general argument. Chapter 1 has outlined these risks. Working on the borderlands of fact and fiction, there is a danger that narrative evidence will be dismissed as merely fiction, rather than understood as 'serious fictions' (Clifford, 1988, p. 100); constructions which are developed in narrative form but firmly based on empirical evidence. Secondly, working most frequently with stories written about teachers but not by them, we run the risk of being seen as hijacking teachers' voices to serve our intentions as authors. Strategies we have used to reduce this risk include routinely sharing the text with teachers described in the text, preferring extended vignettes to decontextualised quotations when citing evidence, and applying a phenomenological test of the authenticity of the narratives. The primary test of authenticity, we think, is that the parts of an ethnographic narrative that apply to an individual participant should be recognisable to the participant and reflect their own language and constructions of reality.

The third risk of working with narrative case studies concerns the uncertainty of interpretation. The act of settling on a final text, of choosing which evidence is shaped into what interpretative web, runs the risk of over-determination. The final text is always just one of the more or less plausible constructions which could have been made of the data as we have understood it. Throughout the book, we have tried to achieve a balance between creating a text that is open to further interpretation, and creating a text that is clear about our interpretation of the events and issues we have described. Strategies we have used to achieve this end include providing multiple interpretations of events and separating lightly interpreted vignettes from more determined sections of interpretation.

Finally, we have approached cautiously the risk of overgeneralisation from cases. The case evidence we have collected, after all, was collected from teachers who were willing to have us in their classrooms — usually for extended periods. Although they include women and men, Australians and Canadians, and elementary and secondary school teachers, they had in common an unusual degree of openness to thinking out loud and sharing their successes and failures with others. As a group, Johanna, David, Mr Ward, Ms Horton, Gerald, Amanda and Geoff were more than usually reflective and collaborative teachers. They may not necessarily be typical of teachers — there are teachers who are less reflective and collaborative — but they are representative of that very large group of teachers who continue to struggle with the goal of reaching high professional standards.

In this final chapter, we return to the theoretical concerns that we have circled around in the first nine chapters, drawing together the evidence from across the case studies of these seven teachers. The chapter is organised in three sections. The first section considers the constitution of teachers' knowledge and the impact of subject matter and milieu on teachers' learning. The second section considers the impact of reflection and collaboration on teachers' learning. The final section considers the implications of these views about teachers' learning for change in the teaching profession.

TEACHERS' LEARNING

For each of the teachers described in this book, much of what they know about teaching is encoded in a set of patterns of practice which structures what they do, day-to-day and moment-to-moment. For Johanna, these patterns of practice included ways of beginning and ending lessons, a preference for introducing class discussions by relating some personal experience of her own, conventions of class discussion designed to involve every child, and sets of tricks developed to manage practical problems. For example, her whole class guitar lessons always involved the same highly individual patterns of practice concerning seating, selection of instruments, distribution of song sheets, identification of the teaching point of the lesson, practice for all students, demonstration by more expert students, and singing through the song as a whole class.

Similarly, the physics lessons taught by David and Mr Ward followed predictable patterns. David's lessons were built around a standard routine. Students began the lessons sitting at their benches. After a few comments from David they collected and assembled the equipment required for the lesson, and then students worked quietly in groups while David circled the room asking questions and coaching students. Few comments were made to the whole class, and often these were organised around student questions. Mr Ward's patterns of practice were different from David's but equally predictable. His most common mode of teaching was to lecture to the class. He rarely asked questions but patiently answered all questions put to him by students. No interaction between students was planned in the lessons, but he was tolerant of discussions between students as they worked on experiments and exercises. One of the few lessons that did not follow this pattern of practice — called a workshop by Mr Ward — involved students working through work sheets or textbook exercises while he stood by to answer individual student questions about the exercises.

Like David and Mr Ward, Amanda and Geoff taught similar content using different patterns of practice. Although they worked closely together and shared a commitment to experiential learning in their grade 1 and 2 classes, they achieved this commitment in different ways. Amanda's lessons used active learning centres and allowed students a good deal of freedom to select and manage their learning. Geoff's lessons were more teacher-centred, relying on his leadership and direction of all activities.

To give a final example, Ms Horton's most familiar pattern of practice involved a cycle of lessons for each new concept. In the first lesson she would introduce the concept through a structured practical activity. While students worked their way through the practical activity she moved around the room, helping students with the activity and asking questions about what they were doing. In the next lesson students would finish writing up the practical activity and Ms Horton would conduct a classroom discussion to consolidate students' learning.

For even a casual observer, the patterns of behaviour of each teacher are easy to establish. What is less obvious is the meaning each of the patterns of practice has for each teacher. Why is Johanna most comfortable when she has her class drawn close together in a circle on the blue oriental rug in the centre of her room? Why does Mr Ward lecture to his class almost all of the time they are in the room, when David teaches the same content but hardly ever talks to the whole class? Why is Gerald able to continue experimenting when he teaches physics when Ms Horton has to return to traditional patterns of teaching in chemistry?

The core of the answer to questions such as these, we think, lies in our conception of teachers' knowledge as a horizon of understanding. Behind the external and observable patterns of each teacher's practice stands a web of tacit and explicit knowledge of what to teach, why it is worth teaching and how it ought to be taught. For each teacher the horizon of understanding is what she or he knows and believes up to now, and the set of content knowledge, patterns of practice and moral predispositions that are carried forward to each new encounter with students. Snapshots of teachers at work emphasise the static, settled aspects of teachers' work: predictable responses to predictable problems. For most teachers, however, teaching is not static. The endless succession of changes in syllabus materials, staffing and scheduling all require changes in teachers' patterns of practice. As teachers encounter each new set of circumstances, their horizons of understanding gradually change. As Gadamer puts it, 'the old and the new continually grow together to make something of living value, without either being explicitly distinguished from the other' (Gadamer, 1975, p. 273).

In many cases, teachers' patterns of practice and their horizons of understanding can be interpreted biographically. Behind the surface features of one way or other of handling a class discussion or structuring a practical session stands a personal conception of what it means to teach, deeply connected to the teachers' own experience as a child, a student and as a beginning teacher.

In Johanna's case, for example, the patterns of practice she used when teaching a whole-class guitar lesson reflect her personal biography and her experience as a teacher. Personally, she was more committed to all students participating in music making than she was concerned that her teaching strategies would hold back gifted or experienced student musicians. Her own experience at being silenced by humiliation when she had tried to learn to play the clarinet had taught her — above all — that she should not discourage students who lacked confidence in music making. Practically, her classroom routines for whole-class guitar teaching had been developed fifteen years before, when she worked as a specialist music teacher and repeated each lesson many times. Through many years of practical experience she had held on to her personal goal that there should be no discouragement, and she retained the settled patterns of practice developed as a beginning teacher. Similarly, Ms Horton's commitment to care for her students reflected her own personal struggle as a school student. She preferred to set hands-on practical work and to move slowly through content to ensure that all children would have an opportunity to learn science concepts. Although this put her in conflict with her head of department, she resisted his attempts to speed up the pace of content coverage by reducing the amount of time allocated to practical work.

Several of the teachers described in this book were under pressure to change as a result of teaching unfamiliar subject matter. For Johanna the need to move outside her areas of expertise in art and music challenged her horizons of understanding. Learning to teach writing posed few difficulties. She was able to adapt existing patterns of practice to the new content area, adapting her over-the-shoulder assistance in art lessons to a process-oriented approach to teaching writing. Her education and her life in the arts community of a big city had led her to value writing alongside other cultural activities, and she was easily able to bridge the small gap in her pedagogical content knowledge. Teaching science to her grade 7 and 8 students opened up much larger gaps in her knowledge. She was sceptical about what counted as knowledge in the school board-provided syllabus. Unaware of the traditional hierarchies of knowledge in science education, she did not appreciate the value or purpose of the recommended content in science — mixtures and solutions. Judging the students to be uninterested in the science content, she was unimpressed with attempts to make the material more interesting to students. After a few weeks of experimentation she dismissed Bill's school science as an attempt to 'bamboozle them into being interested in solutions'. She preferred to choose a topic of more immediate interest to the students, and a long-established pattern of practice. Instead of overcoming the gap between her current patterns of practice and the new content, she moved to content that could be taught from within one of her settled teaching strategies. The students completed a research assignment on health and nutrition and Johanna's horizons of understanding largely remained unchanged by the experience of teaching science.

In the cases of Gerald and Ms Horton, two secondary biology teachers, the uncomfortable experience of teaching grade 10 physics and chemistry had contrasting effects on their horizons of understanding. Gerald felt ill-prepared to teach grade 10 physics. He knew 'the theory', but felt that he lacked 'all the stories, experiences and demonstrations that are so crucial in helping students understand [science]'. In previous years he had been daunted by physics practicals. The story he told in Chapter 6 records his feeling of success in generating an enthusiastic class discussion about Newton's First law of Motion. Emboldened by his success, and undeterred by his head of department's critical comment on the lesson, Gerald decided to continue learning more about teaching physics in an interactive way. He resolved to accept the specific subject-based advice he received, and to look for more opportunities to build on students' understandings of discrepant events.

In contrast, Ms Horton's discomfort in teaching chemical change was increased by the intervention of her head of department. Although Ms Horton was teaching out of her subject specialist area the pressure to do something different came from the context in which she was teaching, not the content. As one of a team of three teaching a grade 10 unit on chemical change — a unit which was used to select students for more advanced science courses in grades 11 and 12 — Ms Horton was under pressure to cover the content at the same pace as the rest of the team. After weeks of being relatively relaxed about time in the classroom, of slowly building students' knowledge through practical activities and discussions, Ms Horton switched to a fast-paced teaching style based on note taking and exercises. She had at first resisted the pressure to switch to this more didactic teaching style, arguing that her students needed more time to consolidate their understanding of the topic. Only when she failed to persuade the head of department to postpone the test did she shift to the traditional teaching style in an attempt 'to cover Avagadro's number, molar calculations and all the equations' in two weeks. Instead of exploring the unfamiliar content using her preferred patterns of practice, guided by her moral disposition to support the weaker students, she was forced to retreat to traditional chalk and talk strategies in order to cover the required content for an across the board test.

In the cases of David and Mr Ward, the differences in the strategies they used to teach the same content might also be explained in terms of the milieus in which they taught. Both Mr Ward and David had long experience of teaching physics in a formal and structured way, with a strong emphasis on speed and accuracy in 'number-plugging' algorithms. They were successful school and university students, graduates of an ·elite university, and practitioners in a content area chosen by students aiming for entrance to the most competitive faculties and universities. For both of them, steeped in the quantitative traditions of physics teaching, the new physics syllabus challenged their existing horizons of understanding. Mr Ward regarded experimentation with a more contextual approach to teaching physics as unsuccessful. After a brief experience with the optics topic, he retreated to his previous patterns of practice. David's reading of his experimentation with teaching optics was quite different. Instead of retreating to a quantitative approach when his students' test results were disappointing, he resolved to develop more contextual assessment practices.

There may be biographical differences behind these responses to essentially similar events, but there were also important differences in the milieu in which each teacher was working. The highly motivated students in Mr Ward's class, intent on the opportunities that success in physics might open up for them at university, were unwilling to tolerate poor test results. For them, their parents and other staff at their private girl's school, academic results were more important than educational ideology. Mr Ward's reading of his milieu was that it allowed him very little latitude for experimentation. For David, the head of department in his working-class school and the only physics teacher, there seemed to be little pressure against experimentation. He believed deeply in the value of more contextual physics teaching and was unimpeded by the milieu of the school from expanding his knowledge of teaching the new syllabus.

Similarly, there was no pressure from the milieu on Johanna's choices in teaching grade 7 and 8 science. There was no specialist science teacher on the staff, there were no curriculum controls in the school to ensure that she followed the school board guidelines, and the parents had chosen the school on the basis of its alternative teaching practices. In contrast with Ms Horton, Johanna was free to pursue her own version of science teaching without the intrusion of a head of science, the scrutiny of a group of other teachers working on the same topic at the same time, or the pressure to prepare her students for a high stakes testing program.

Much of what teachers know and are able to do may be explained by the resilience of established patterns of practice, the interaction of teachers' biography and experience, and the pressures of subject matter and milieu. Teachers' knowledge of their own teaching, however, is always incomplete in two major ways. In principle, it is incomplete because all understanding is incomplete. As we have characterised it, teachers' knowledge is constantly but often imperceptibly changing. Each new encounter with experience allows the possibility of new horizons of understanding. The second way in which teachers' understanding of teaching is incomplete is that it is personal knowledge, knowledge used by an individual to negotiate the world as he or she experiences it. This experiential world is different in crucial ways from the world experienced by other teachers working in similar milieus, teachers working in the same milieu, and from students in their classes.

Teachers do not have access to students' experience of their classes; they have access only to what they take to be students' reactions to their teaching. As the student perspectives of Ms Horton's lessons showed, Punipa and Karl inhabit different interpretive spaces. They understand events differently from each other, and differently from Ms Horton. A focus on the teacher's taken-for-granted interpretation of teaching the chemical change topic yields a story about the struggle between Ms Horton's ethic of care and her head of department's ethic of responsibility. Ms Horton's story plays out this struggle, showing the patterns of practice that she follows when she is able to teach as she prefers, and the patterns of practice she is forced to adopt in order to fit into her head of department's plans. Students' experience in the classroom is affected by Ms Horton's variation in her patterns of practice, but is not determined by her action. For Punipa, highly motivated to academic success and uncomfortable with Ms Horton's view that 'there are no right or wrong answers', the shift from informal practical lessons and class discussions to chalk and talk was beneficial. For reasons that may be linked to her cultural background, she was more comfortable with the clarity, pace and explicitness of the lessons at the end of the unit on chemical change. Punipa's particular needs were hidden from view and unconsidered in the pedagogical struggle between Ms Horton and her head of department, but Punipa's clear preference was for lessons where 'the teacher is in charge of the whole thing'.

Karl's reading of Ms Horton's lessons was quite different from Punipa's. He appreciated Ms Horton's friendly approach — 'She sort of smiles and says hello and stuff like that' — and preferred practical lessons to lessons where 'you're just writing down everything from the board'. He was willing enough to participate in class activities, but easily discouraged by content that seemed too complex, activities that seemed boring, or instructions that had too many steps. Whatever patterns of teaching Ms Horton adopted, Karl's behaviour was as likely to be influenced by the reaction of his friends as by any change in his teacher's patterns of practice.

It is not only students' alternative perspectives that are occluded from the view of individual teachers. Teachers' typically work in isolation, unaware of the learning of teachers around them, and with limited opportunities for reflection on their own practice. In the next section, we summarise the arguments we have made about the importance of reflection and collaboration in promoting teachers' learning.

REFLECTION AND COLLABORATION

Most of the stories of teachers' learning that we have included in this book are stories of teachers' learning through reflection. Gerald's account of his attempts to teach grade 10 physics provides several layers of commentary on his teaching, including his own reflections about his learning from reading the reactions of his head of department and a student teacher. David and Mr Ward talked to us about their attempts to blend the old and the new in grade 11 physics teaching. Ms Horton talked to us about her struggle to maintain her ethic of care for less academic students. Amanda and Geoff, and Johanna and Bill constructed working relationships that increased the space available for reflection on their teaching practice.

The range of reflective practices across these cases is very wide. In Chapter 7 we argued that reflection serves many interests and may take many forms. Using the case study material concerning Johanna, we distinguished among technical, personal, problem-solving and critical interests in reflection. Technical interests concern fidelity to an externally mandated model of teaching. Reflection with a personal interest involves reflection on the relationship between a teacher's life and work. Problem solving concerns the resolution of the problems of professional action. Reflection with a critical interest involves explicit questioning of taken-for-granted thoughts, feelings and actions, usually with reference to some explicit theoretical position. Of these four interests, Johanna most often reflected with a personal interest, an interest in connecting teaching experience with understanding of her own life. When Mr Ward decided to abandon contextual teaching of the grade 11 physics syllabus, it was as a result of reflection with a technical interest. Which method, he had asked himself, was the most efficient way of preparing for the external examination. When Gerald was working through the implication of his grade 10 physics teaching, he did so with a problem solving interest: how could he get beyond the 'trolleys, ticker timers and light boxes' to physics practicals which challenged students' understanding? When Ms Horton was struggling with the implications of the common test in grade 10 chemistry, she was reflecting with a critical interest, arguing in terms of social theory that her head of department was 'authoritarian' and 'using male domination tactics' to ensure her conformity to his wishes.

An alternative way of conceptualising the range of reflective practices, also introduced in Chapter 7, is in terms of four forms of reflection identified as introspection, replay and rehearsal, inquiry and spontaneity. Introspection, a conscious form of reflection conducted at some distance from the action was displayed by Johanna in her account of the relationship between her own experiences in 'druggie times' of the Sixties and her belief in encouraging her students to take responsibility for their decision making. Gerald's cycle of writing about classroom events may be characterised as replay and rehearsal. It took place after a lesson, involved replaying the events through the eyes of several commentators, and was concluded by his own reflection on what he had learned from the lesson and the commentaries. David's cycle of investigation into ways of teaching grade 11 physics may be characterised as inquiry, a process of deliberate movement from planning to make some changes, teaching differently, evaluating the impact of the changes, and then planning another round of teaching based on the evaluation. Finally, the case evidence presented in this book provides many examples of reflection as spontaneity. This form of reflection takes place in the moment, as teachers read the classroom action and, without conscious thought, move from one activity structure to another. Like Schön's 'jazz-player' form of reflection in action (1987), Johanna improvised a role-play, Bill replaced a small group discussion with a demonstration, and David departed from his lesson plan to explore Michael's non-standard observation of total internal reflection.

With perhaps the exception of the form of reflection we have called spontaneity, all of these forms and interests for reflection are supported through teachers' collaboration with their colleagues. In Chapter 8 we have identified a series of qualities of effective collaborative partnerships among teachers. Effective partnerships often involve teachers from similar traditions of teaching, with similar hopes and dreams. Amanda and Geoff, and Johanna and Bill shared sufficiently similar understandings of teaching that they could work together to solve classroom problems. Although Johanna and Ms Horton had never met, their commitment to teach in ways that minimised the possibility of discouragement of students would have provided a basis for collaboration. Differences also create opportunities for collaboration. Part of the success of Johanna and Bill's collaboration was that they shared values but came from different discipline backgrounds. Similarly, although David and Mr Ward were not working together, the combination of their similar backgrounds in quantitative approaches to teaching physics and their different levels of commitment to experimenting with contextual approaches to physics would have provided a basis of shared understandings and some reasons to learn from each other.

Collaborative relationships are sites of power, and contain the possibility that collaboration may give way to competition. We have identified several practical steps teachers and researchers can take to defuse the force of power in collaborative relationships, and several moral dispositions which limit the likelihood of power being misused. Among the practical issues are symmetry, risk sharing and fair exchange. Amanda and Geoff's relationship was symmetrical because they were both new teachers making their way in the profession. Johanna and Bill's relationship involved different but equal roles. In the latter case, the teacher's willingness to take on the role of the researcher and the researcher's willingness to take on the role of the teacher reduced the risk that either partner would fail to understand the role of the other partner. The third practical quality of effective collaboration, especially important for teacher-researcher collaboration, is the willingness to share the effort of maintaining the collaboration. Often this involves sharing the workload; for researchers it may involve contributing practical labour — planning, marking, teaching — in return for the time teacher-researcher collaboration costs a teacher.

In addition to these practical qualities of collaboration, Chapter 7 identified several crucial moral dispositions: a commitment to emergence, trust and humility. Unlike the contrived collegiality (Hargreaves, 1994) of performance management programs, the collaborations described in this book were based on open-ended agendas, allowing issues to emerge and be taken up as part of extended personal relationships. Without a commitment to trust between the partners, it is not possible for teachers to take the risks that lead to new learning about teaching. Finally, without a commitment to humility, it is difficult for teachers to establish a proper balance between the professional pride that leads teachers to want to learn more, and the professional modesty that allows teachers to admit that there is much more that they would want to do.

When researchers are collaborative partners, as they will normally be in research requiring extended periods of participation in classrooms, there are some particular ethical issues. Teaching is such a personal act that writing about teaching needs to be done with a concern for sustaining human relationships. As the stories we have told about Malcolm and Simon show, researchers and teachers may easily misunderstand each other — especially when the text paints a portrait that the teacher does not recognise. For this reason, it is necessary for narrative researchers to adopt a relational approach to ethics (Noddings, 1988). Uniform application of law-like general rules, which is normally preferred by university ethics committees, does not deal adequately with ethical issues arising in the context of what must inevitably be personal relationships between teachers and researchers. These issues are more sharply felt when there are ideological differences between teachers and researchers. For Johanna and Bill, there were few differences. But there is a greater challenge in working with teachers such as Malcolm, Simon and Mr Ward, where there were clear ideological differences. As Hargreaves (1996) has argued, there is a risk if narrative researchers attempt to build a body of knowledge based only on working with teachers who share their literary, humanistic, constructivist preconceptions. Collaboration between teachers and researchers is essential to narrative research, and it must include respectful and ethical collaboration with the whole range of teachers, including teachers whose preconceptions about teaching researchers do not share.

Beginning with stories about Johanna, David, Mr Ward, Ms Horton, Gerald, Amanda, Geoff and others, this book has developed an argument about the construction of teachers' knowledge. This final chapter has drawn together the threads of the argument. What we believe we have learned about teachers' learning can be summarised in the following nine propositions.

1. *Patterns of practice.* Much of what teachers know about teaching is encoded in the patterns of practice that structure what they do, day-to-day and moment-to-moment. Behind these patterns of practice stands a web of tacit and explicit knowledge of what to teach, why it is worth teaching and how it is to be taught.

2. *Horizons of understanding.* This web of content knowledge, patterns of practice and moral predispositions forms a horizon of understanding — what each teacher knows, believes and is able to do up to now.

3. *Biography and experience.* Teachers' horizons of understanding reflect their biography and experience as a child, a student and as a teacher.

4. *Subject matter.* Changes in teaching content and context create gaps in teachers' understanding. Sometimes teachers overcome these gaps and their horizons of understanding expand. For other teachers and at other times the challenge of unfamiliar content

leads to a retreat or consolidation of horizons of understanding.

5. *Milieu*. The milieu in which teachers work may facilitate or constrain teachers' capacity to overcome gaps in their understanding.

6. *Incompleteness*. Teachers' understanding is always incomplete. It is incomplete in principle because each new encounter with experience contains the possibility of growth and change. Teachers' understanding is also incomplete because it is personal knowledge, circumscribed by teachers' personal experience and occluding the perspectives students and other teachers have of the same experience.

7. *Reflection*. Reflection is a powerful force in opening up the possibility of change. It takes many forms, ranging from spontaneous learning in the midst of action to introspection at some distance from the classroom. Reflection may also have a range of purposes, depending on whether teachers' interests in reflection are technical, personal, practical or critical.

8. *Collaboration*. Collaborative partnerships among teachers or between teachers and researchers open up the possibility of change. The qualities of successful collaborative partnerships include symmetry, sharing the risk, fair exchange, a commitment to emergence, trust and humility.

9. *Ethics*. Collaboration between teachers and researchers exposes some ethical issues. If both teachers and researchers are to learn from collaborative partnerships, researchers must balance their theoretical obligation to work with teachers whose views they do not share, with their personal obligation to respect the perspectives of their collaborative partners.

THE POSSIBILITY OF CHANGE IN SCIENCE EDUCATION

Science education may be characterised as having a Whig view of history — a view that history is progress, that human beings and institutions are moving on an upward path towards a perfectible future, and that each succeeding year will bring improvements over the last year. This view, which stands behind the reports and reviews and commissions of excellence which have shaped public debate about science education in the OECD countries (Wallace & Louden, 1998), is somewhat at odds with what we have written about teachers' learning in science education. We have argued that what teachers do depends on their biography and experience, that they search for and find it hard to shift from settled patterns of practice, that their knowledge develops gradually and hesitantly, that subject matter knowledge and milieu constrain teachers' capacity to change, and that experimentation is often followed by retreat to established practices.

Despite these pressures for continuity, opportunities for reflection and collaboration open up the possibilities for change. Sometimes gaps in content knowledge close down these possibilities, but Johanna's retreat from school board science contrasts with Gerald's commitment to continue to experiment with his physics teaching. Sometimes school structures and cultures close down the possibilities for change, but Mr Ward's retreat from contextual physics teaching contrasts with David's commitment to review his assessment strategies. Teachers' knowledge is always incomplete, always in the process of becoming. By continuing to reflect on their practice and work in collaborative partnerships with teachers or researchers, teachers such as Johanna, Amanda, Geoff, David and Gerald maximised the possibility that they would overcome gaps in their knowledge or constraints imposed by the cultures and structures within which they worked.

Proposals for reform, frequently couched in the form of a national crisis in science education, will fail if they do not take account of the ways in which teachers' knowledge grows and changes. The irony is that many contemporary reforms in science education focus on teachers being more accepting of children's constructions of reality. By encouraging students to make their own ideas explicit, presenting them with discrepant events which challenge these ideas, and providing students with opportunities to use new ideas in a range of situations, students' understanding of science concepts is expected to grow. This constructivist vision of children's learning parallels our account of teachers' learning. For teachers to change — to become more constructivist or contextual or to teach for understanding (Louden & Wallace, 1995) — then their knowledge of these new ways of teaching must be allowed to grow through gradual reconstruction of their current understandings of teaching. For this reason it is crucial that reformers do not fall into what we have called the constructivist paradox (Louden & Wallace, 1994), being critical of the incompleteness or wrong-headedness of teachers especially when they are regarded as wrong-headed about constructivist approaches to science. Teachers do not understand by being told, any more than children understand by being told. There is a striking epistemological parallel between teachers' learning and students' learning. In each case, fundamental changes in understanding require that teachers or reformers begin with respect for what people already take to be true.

This is a hard lesson for Whiggish reformers to learn. It has always been easier for reformers to secure funding for the development stage of reform projects than the implementation stage. Too often, funds and energy are exhausted by the time the early adopters have implemented a reform, and long before the more cautious and traditional teachers have had an opportunity to consider and experiment with new approaches and materials. Because teachers' prior understandings are the bedrock on which a reform must be built, reformers need to approach teachers with humility, tact and respect for tradition. They need also to provide opportunities for reflection and collaboration among teachers. Without the enthusiasm of reform-minded teachers, researchers and school board officials, there would be no change. More importantly, few reforms are likely to be sustained without respect for teachers' past and present understanding of their work.

REFERENCES

Apple, M. (1990). *Ideology and the curriculum*, New York, Routledge.

Ballard, B. (1989). 'Overseas students and Australian academics: Learning and teaching styles', in B. Williams (ed.), *Overseas students in Australia*, Canberra, Australia, International Development Program of Australian Universities and Colleges, 87–98.

Barone, T. (1995). 'Persuasive writings, vigilant readings, and reconstructed characters: The paradox of trust in educational storysharing', *Qualitative Studies in Education*, *8*(1), 63–74.

Barton, A.C. (1998). *Feminist science education*, New York, Teachers College Press.

Berlack, A. & Berlack, H. (1981). *Dilemmas of teaching*, New York, Methuen.

Biggs, J. (1991). 'Approaches to learning in secondary and tertiary students in Hong Kong: Some comparative studies', *Educational Research Journal*, *6*, 27–39.

Blythe, W.A.L. (1965). *English primary education: A sociological description — Vol. 2. Background*, London, Routledge and Kegan Paul.

Bolster, A.J. (1983). 'Towards a more effective model of research on teaching', *Harvard Educational Review*, *53*, 294–308.

Brickhouse, N. (1992). 'Ethics in field-based research: Ethical principles and relational considerations', *Science Education*, *76*(1), 93–103.

Britzman, D.P. (1991). *Practice makes practice: A critical study of learning to teach*, Albany, NY, SUNY Press.

Brodkey, L. (1987). 'Writing critical ethnographic narratives', *Anthropology and Education Quarterly*, *18*, 67–76.

Bruner, J. (1986). *Actual minds, possible worlds*, Cambridge, MA, Harvard University Press.

Butt, R.L. & Raymond, D. (1987). 'Arguments for using qualitative approaches in understanding teacher thinking: The case for biography', *Journal of Curriculum Studies*, *7*(1), 62–93.

Carr, W. & Kemmis, S. (1986). *Becoming critical: Education, knowledge and action research*, London, Falmer Press.

Carter, K. (1992). 'Creating cases for the development of teacher knowledge', in T. Russell & H. Munby (eds.), *Teachers and teaching: From classroom to reflection*, London, Falmer Press, 109–123.

Carter, K. (1993). 'The place of story in the study of teaching and teacher education', *Educational Researcher*, *22*(1), 5–12.

Christiansen, H., Goulet, L., Krentz, C. & Maeers, M. (eds.). (1997). *Recreating relationships: Collaboration and educational reform*, Albany, NY, SUNY Press.

Cizek, G. (1995). 'Crunchy granola and the hegemony of narrative', *Educational Researcher*, *24*(2), 26–30.

Clandinin, D.J. & Connelly, F.M. (1988). 'Studying teachers' knowledge of classrooms: Collaborative research, ethics, and the negotiation of narrative', *The Journal of Educational Thought*, 22(2A), 269–293.

Clandinin, D.J. & Connelly, F.M. (1991). 'Narrative and story in practice and research', in D.A. Schön (ed.), *The reflective turn: Case studies in and on educational practice*, New York, Teachers College Press, 174–198.

Clandinin, D.J. & Connelly, F.M. (1992). 'Teacher as curriculum maker', in P.W. Jackson (ed.), *Handbook of research on curriculum*, New York, Macmillan, 363–401.

Clandinin, D.J. & Connelly, F.M. (1994). 'Personal experience methods', in N.K. Denzin & Y.S. Lincoln (eds.), *Handbook of qualitative research*, Thousand Oaks, CA, Sage, 413–427.

Clandinin, D.J. & Connelly, F.M. (eds.). (1995). *Teachers' professional knowledge landscapes*, New York, Teachers College Press.

Clandinin, D.J. & Connelly, F.M. (1996). 'Teachers' professional knowledge landscapes: Teacher stories — stories of teachers — school stories — stories of school', *Educational Researcher*, 25(3), 24–30.

Clandinin, D.J. & Connelly, F.M. (1997). 'Asking questions about telling stories', in C. Kridel (ed.), *Writing educational biography: Explorations in qualitative research*, New York, Garland, 202–209.

Clandinin, D.J., Davies, A., Hogan, P. & Kennard, B. (eds.). (1993). *Learning to teach: Teaching to learn,* New York, Teachers College Press.

Clifford, J. (1988). *The predicament of culture: Twentieth-century ethnography, literature, and art*, Cambridge, MA, Harvard University Press.

Clift, R.T., Houston, W.R. & Pugach, M.C. (eds.). (1990). *Encouraging reflective practice in education*, New York, Teachers College Press.

Cochran-Smith, M. & Lytle, S.L. (1993). *Inside/outside: Teacher research and knowledge,* New York, Teachers College Press.

Connelly, F.M. & Clandinin, D.J. (1985). 'Personal practical knowledge and the modes of knowing: Relevance for teaching and learning', in E. Eisner (ed.), *Learning and teaching the ways of knowing* (84th Yearbook of the National Society for the Study of Education), Chicago, IL, University of Chicago Press, 174–198.

Connelly, F.M. & Clandinin, D.J. (1988). *Teachers as curriculum planners: Narratives of experience*, Toronto, Canada, OISE Press & New York, Teachers College Press.

Connelly, F.M. & Clandinin, D.J. (1990). 'Stories of experience and narrative inquiry', *Educational Researcher*, 19(5), 2–14.

Costa, V.B. (1993). 'School science as a rite of passage: A new frame for familiar problems', *Journal of Research in Science Teaching*, 30(7), 649–668.

Costa, V.B. (1997). 'Honours chemistry: High-status knowledge or knowledge about high status', *Journal of Curriculum Studies*, 29(3), 289–313.

Cuban, L. (1992). 'Managing dilemmas while building professional communities', *Educational Researcher*, 21(1), 4–11.

Demidenko, H. (1995). *The hand that signed the paper*, Sydney, Australia, Allen & Unwin.

Denzin, N.K. & Lincoln, Y.S. (eds.). (1994). *Handbook of qualitative research*,

Thousand Oaks, CA, Sage.

Dewey, J. (1910/1933). *How we think*, Lexington, MA, D.C. Heath & Co.

Dewey, J. (1958). *Experience and nature*, New York, Dover.

Dixon, G. (1996). 'The Mudrooroo dilemma', *Westerly*, Spring, 5–6.

Eisner, E. (1997). 'The promise and perils of alternative forms of data representation', *Educational Researcher*, 26(6), 4–10.

Elbaz, F.L. (1983). *Teacher thinking: A study of practical knowledge*, London, Croom Helm.

Ellsworth, E. (1992). 'Why doesn't this feel empowering? Working through the repressive myths of critical pedagogy', in C. Luke & J. Gore (eds.), *Feminisms and critical pedagogy*, New York, Routledge, 90–119.

Fairclough, N. (1992). *Discourse and social change*, London, Polity Press.

Feiman-Nemser, S. & Floden, R. (1986). 'The cultures of teaching', in M.C. Wittrock (ed.), *Handbook of research on teaching*, New York, Macmillan, 505–526.

Fenstermacher, G.D. (1994). 'The knower and the known: The nature of knowledge in research on teaching', *Review of Research in Education*, 20, 3–56.

Feyerabend, P. (1975). *Against method: Outline of a theory of knowledge*, London, Verso.

Feyerabend, P. (1995). *Killing time*, Chicago, IL, University of Chicago Press.

Foucault, M. (1972). *The archaeology of knowledge* (A. Sheridan, Trans.), New York, Pantheon.

Foucault, M. (1977). *Discipline and punish: The birth of the prison* (A. Sheridan, Trans.), London, Penguin.

Fullan, M. (1991). *The new meaning of educational change*, London, Cassell.

Fullan, M. (1993). *Change forces: Probing the depths of educational reform*, London, Falmer Press.

Gadamer, H.-G. (1975). *Truth and method* (G. Barden & J. Cumming, Edited and Trans.), New York, Seabury Press.

Garner, H. (1995). *The first stone*, Sydney, Australia, Picador.

Geertz, C. (1988). *Works and lives: The anthropologist as author*, Stanford, CA, Stanford University Press.

Gilligan, C. (1982). *In a different voice*, Cambridge, MA, Harvard University Press.

Glaser, B.G. & Strauss, A.L. (1967). *The discovery of grounded theory: Strategies for qualitative research*, Chicago, IL, Aldine.

Goetz, J. & LeCompte, M.D. (1984). *Ethnography and qualitative design in educational research*, Orlando, FL, Academic Press.

Goodson, I.F. (ed) (1992). *Studying teachers' lives*, New York, Teachers College Press.

Goodson, I.F. (1995). 'The story so far: Personal knowledge and the political', in J.A. Hatch & R. Wisniewski (eds.), *Life history and narrative*, London, Falmer Press, 89–98.

Goodson, I.F. & Marsh, C.J. (1996). *Studying school subjects: A guide*, London, Falmer Press.

Gordon, T. (1970). *Parent Effectiveness Training: The 'no-lose' program for raising responsible children*, New York, P.H. Wyden.

Gordon, T. (1974). *TET: Teacher Effectiveness Training*, New York, P.H. Wyden.

Gramsci, A. (1971). *Selections from the prison notebooks of Antonio Gramsci* (J. Quintin-Hoare & G. Nowell Smith, Edited and Trans.), London, Lawrence & Wishart.

Greenblatt, R. (1991). *Marvellous possessions: The wonder of the new* world, Oxford, UK, Clarendon Press.

Grossman, P.L. & Stodolsky, S.S. (1994). 'Considerations of content and the circumstances of secondary school teaching', *Review of Research in Education, 20*, 179–221.

Grumet, M. (1978). Curriculum as theatre: Merely players. *Curriculum Inquiry, 8*, 37–62.

Grundy, S. (1987). *Curriculum: Product or praxis*. London: Falmer Press.

Guba. E.G. (1978). *Toward a methodology of naturalistic inquiry in educational evaluation*, Los Angeles, CA, UCLA Center for the Study of Evaluation.

Guba, E.G. & Lincoln, Y.S. (1981). *Effective evaluation*, San Francisco, CA, Jossey Bass.

Guild, P. (1994). 'The culture/learning style connection', *Educational Leadership, 51*(8), 16–21.

Gunstone, R. (1990). '"Children's science": A decade of developments in constructivist views of science teaching and learning', *Australian Science Teachers Journal, 36*(4), 9–19.

Gunstone, R. (1992). 'Constructivism and metacognition: Theoretical issues and classroom studies', in R. Duit, F. Goldberg & H. Niedderer (eds.), *Research in physics learning: Theoretical issues and empirical studies*, Kiel, Germany, University of Kiel, Institute for Science Education, 129–140.

Habermas, J. (1971). *Knowledge and human interests* (J. Shapiro, Trans.), Boston, MA, Beacon Press.

Hargreaves, A. (1994). *Changing teachers, changing times: Teachers' work and culture in the postmodern age*, London, Cassell.

Hargreaves, A. (1996). 'Revisiting voice', *Educational Researcher, 25*(1), 12–19.

Hatch J.A. & Wisniewski, R. (1995). 'Life history and narrative: Questions, issues, and exemplary works', in J.A. Hatch & R. Wisniewski (eds.), *Life history and narrative*, London, Falmer Press, 113–133.

Hollingsworth, S. (1992). 'Learning to teach through collaborative conversation', *American Educational Research Journal, 29*(2), 373–404.

Holmwood, J. (1996). 'Abject theory', *Australian and New Zealand Journal of Sociology, 32*(2), 86–108.

Huberman, M. (1983). 'Recipes for busy kitchens', *Knowledge: Creation, Diffusion, Utilization, 4*(4), 478–510.

Huberman, M. (1992). 'Teacher development and instructional mastery', in A. Hargreaves & M.G. Fullan (eds.), *Understanding teacher development*, London, Cassell and New York: Teachers College Press, 122–142.

Jackson, M. (1995). *At home in the world*, London, Harper Perennial.

Jackson, P.W. (1968). *Life in classrooms*, New York, Holt, Rinehart & Winston.

Kuhn, T. (1962). *The structure of scientific revolutions*, Chicago, IL, University of Chicago Press.

Lampert, M. (1984). 'Teaching about thinking and thinking about teaching', *Journal of Curriculum Studies*, *16*(1), 1–16.

Lather, P. (1991). *Getting smart: Feminist research and pedagogy with/in the postmodern*, New York, Routledge.

Latour, B. (1987). *Science in action*, Cambridge, MA, Harvard University Press.

Laurie, V. (1996, July 20–21). 'Identity crisis', *Weekend Australian Magazine*, pp. 28–32.

LeCompte, M.D., Millroy, W.L. & Preissle, J. (eds.). (1992). *The handbook of qualitative research in education*, New York, Academic Press.

Lincoln, T.S. (1990). 'Towards a categorical imperative for qualitative research', in E.W. Eisner & A. Peshkin (eds.), *Qualitative inquiry in education: The continuing debate*, New York, Teachers College Press, 277–295.

Lincoln, Y.S. & Denzin, N.K. (1994). 'The fifth moment', in N.K. Denzin & Y.S. Lincoln (eds.), *Handbook of qualitative research*, Thousand Oaks, CA, Sage, 575–586.

Lincoln, Y.S. & Guba, E.G. (1985). *Naturalistic inquiry*, Beverley Hills, CA, Sage.

Lincoln, Y.S. & Guba, E.G. (1989). 'Ethics: The failure of positivist science', *Review of Higher Education*, *12*(3), 221–240.

Longley, K.O. (1997). 'Fabricating otherness: Demidenko and exoticism', *Westerly*, Autumn, 29–45.

Lortie, D.C. (1975). *Schoolteacher: A sociological study*, Chicago, IL, University of Chicago Press.

Louden, W. (1991). *Understanding teaching: Continuity and change in teachers' knowledge*, London, Cassell and New York, Teachers College Press.

Louden, W. & Wallace, J. (1993). 'Competency standards in teaching: Exploring the case', *Unicorn*, *19*(1), 45–53.

Louden, W. & Wallace, J. (1994). 'Knowing and teaching science: The constructivist paradox', *International Journal of Science Education*, *16*(6), 649–657.

Louden, W. & Wallace, J. (1995, April). 'What we don't understand about teaching for understanding', paper presented at the annual meeting of the National Association for Research in Science Teaching, San Francisco, CA.

Louden, W. & Wallace, J. (1996). *Quality in the classroom: Learning about teaching through case studies*, Sydney, Australia, Hodder Education.

Loving, C.C. (1997). 'From the summit of truth to its slippery slopes: Science education's journey through positivist-postmodern territory', *American Educational Research Journal*, *34*(3), 421–452.

Lyotard, J.-F. (1984). *The postmodern condition: A report on knowledge* (G. Bennington, Trans.), Manchester, UK, Manchester University Press.

Martin, J.R. (1990). 'Literacy in science: Learning to handle text as technology', in F. Christie (ed.), *Literacy for a changing world*, Hawthorn, Australia, Australian Council for Educational Research, 79–117.

McAlpine, L. & Crago, M. (1997). 'Who's important . . . here anyway?: Co-constructing research across cultures', in M. Christiansen, L. Goulet, C. Krentz & M. Maeers (eds.), *Recreating relationships: Collaboration and educational reform*, Albany, NY, SUNY Press, 105–114.

McIntyre, A. (1981). *After virtue: A study in moral theory*, Notre Dame, IN, University of Notre Dame Press.

McLaughlin, M. (1993). 'What matters most in teachers' workplace?', in J.W. Little & M. McLaughlin (eds.), *Teachers' work: Individuals, colleagues and contexts*, New York, Teachers College Press, 79–103.

McWhorter, P., Jarrard, B., Rhoades, M., Tatum, B. & Wiltcher, B. (1998). 'The importance of research relationships, the power of a research community', in B.S. Bisplinghoff & J.B. Allen (eds.), *Engaging teachers: Creating teaching and researching relationships*, Portsmouth, NH, Heinemann, 43–52.

Miles, M.B. & Huberman, A.M. (1993). *Qualitative data analysis: A sourcebook of new methods* (2nd ed.), Newbury Park, CA, Sage.

Miller, J.L. (1990). *Creating spaces and finding voices: Teachers collaborating for empowerment*, Albany, NY, SUNY Press.

Miller, J.L. (1992). 'Exploring power and authority issues in a collaborative research project', *Theory into Practice, 31*(2), 165–172.

Munby, H. & Russell, T. (1993, April). *The authority of experience in learning to teach: Messages from a physics methods class*, paper presented at the annual meeting of the American Educational Research Association, Atlanta, GA.

Munby, H. & Russell, T. (1995). 'Towards rigour with relevance: How can teachers and teacher educators claim to know', in T. Russell & F. Korthagen (eds.), *Teachers who teach teachers: Reflections on teacher education*, London, Falmer Press, 172–184.

Neill, R. (1995, 24 August). 'Demidenko: Life fiction goes too far', *The Australian*, p. 13.

Nias, J. (1989). *Primary teachers talking*, London, Routledge & Kegan Paul.

Nias, J., Southworth, G. & Yeomans, R. (1989). *Staff relationships in the primary school: A study of organisational cultures*, London, Cassell.

Noddings, N. (1984). *Caring: A feminine approach to ethics and moral education*, Berkeley, CA, University of California Press.

Noddings, N. (1988). 'An ethic of caring and its implications for instructional arrangements', *American Journal of Education, 96*(2), 215–230.

Olson, J.K. (1985). 'Changing our ideas about change', *Canadian Journal of Education, 10*(3), 294–308.

Ondaatje, M. (1987). *In the skin of a lion*, New York, Knopf.

Paley, V.G. (1989). *White teacher*, Cambridge, MA, Harvard University Press.

Parker, L.H., Wallace, J. & Fraser, B.J. (1993). 'The renewal of science teachers' knowledge: A pilot professional development project', *South Pacific Journal of Teacher Education, 21*(2), 169–177.

Parker, L.H., Wildy, H., Wallace, J. & Rennie, L. (1994). *Post-compulsory schooling Physics implementation: Issues in classroom teaching and assessment. Evaluation Report Number 3* (Report for the Education Department of Western Australia), Perth, Australia, Key Centre for School Science and Mathematics, Curtin University of Technology.

Phillips, D.C. (1997). 'Telling the truth about stories', *Teaching and Teacher Education, 13*(1), 101–109.

Polkinghorne, D.E. (1995). 'Narrative configuration in qualitative analysis', *Qualitative Studies in Education*, 8(1), 5–23.

Polkinghorne, D.E. (1997). 'Reporting qualitative research as practice', in W.E. Tierney & Y.S. Lincoln (eds.), *Representation and the text: Reframing the narrative voice*, Albany, NY, SUNY Press, 3–21.

Punch, M. (1994). 'Politics and ethics in qualitative research', in N.K. Denzin & Y.S. Lincoln (eds.), *Handbook of qualitative research*, Thousand Oaks, CA, Sage, 83–96.

Pybus, C. (1995). 'Cassandra Pybus reviews Helen Garner's "The first stone"', *Australian Book Review*, May, 6–8.

Russell, T. & Munby, H. (eds.). (1992). *Teachers and teaching: From classroom to reflection*, London, Falmer Press.

Said, E.W. (1993). *Culture and imperialism*, London, Chatto & Windus.

Said, E.W. (1995). *Orientalism: Western conceptions of the Orient*, London, Penguin.

Schama, S. (1989). *Citizens: A chronicle of the French Revolution*, London, Penguin.

Schama, S. (1991). *Dead certainties: Unwarranted speculations*, London, Granta.

Schön, D.A. (1983). *The reflective practitioner*, New York, Basic Books.

Schön, D.A. (1987). *Educating the reflective practitioner*, San Francisco, CA, Jossey-Bass.

Schroeder, D. & Webb, K. (1997). 'Between two worlds: University expectations and collaborative research realities', in M. Christiansen, L. Goulet, C. Krentz & M. Maeers (eds.), *Recreating relationships: Collaboration and educational reform*, Albany, NY, SUNY Press, 233–246.

Schulz, R. (1997). *Interpreting teacher practice ... Two continuing stories*, New York, Teachers College Press.

Schutz, A. & Luckmann, T. (1973). *Structures of the life-world*, Evanston, IL, Northwestern University Press.

Sharp, R. & Green, S. (1975). *Education and social control: A study in progressive education*, London, Routledge & Kegan Paul.

Shulman, J.H. (1992). *Case methods in teacher education*, New York, Teachers College Press.

Shulman, L.S. (1987). 'Knowledge and teaching: Foundations of the new reform', *Harvard Educational Review*, 57(1), 1–22.

Shulman, L.S. (1992). 'Towards a pedagogy of cases', in J.H. Shulman (ed.), *Case methods in teacher education*, New York, Teachers College Press, 21.

Siskin, L.S. (1994). *Realms of knowledge: Academic departments in secondary schools*, London, Falmer Press.

Smith, L. (1990). 'Ethics in qualitative field research: An individual's perspective', in E. Eisner & A. Peshkin (eds.), *Qualitative inquiry in education: The continuing debate*, New York, Teachers College Press, 258–276.

Snow, C.P. (1964). *The two cultures: And a second look* (2nd ed.), Cambridge, UK, Cambridge University Press.

Stake, R. (1988). 'Case study methods in educational research: Seeking sweet water', in R.M. Jaeger (ed.), *Complementary methods for research in education*, Washington, DC, American Educational Research Association, 253–265.

Stevenson, H.W. & Lee, S.-Y. (1990). 'Contexts of achievement: A study of American, Chinese and Japanese children', *Monographs of the Society for Research in Child Development*, *55*(1–2, Serial No. 221).

Sykes, G. & Bird, T. (1992). 'Teacher education and the case idea', *Review of Research in Education*, *18*, 457–521.

Trinh, T. Minh-ha. (1989). *Woman, native, other*, Bloomington, IN and Indianapolis, IN, Indiana University Press.

Tripp, D. (1993). *Critical incidents in teaching: Developing professional judgement*, London, Routledge.

Van Manen, M. (1990). *Researching lived experience: Human science for an action sensitive pedagogy*, Albany, NY, SUNY Press.

Wallace, J. (1998). 'Collegiality and teachers' work in the context of peer supervision', *The Elementary School Journal*, *99*(1), 87–98.

Wallace, J. & Chou, C.-Y. (1998, April). 'Essence and variation: The meaning of student cooperation in Taiwanese and Australian science classrooms', paper presented at the annual meeting of the National Association for Research in Science Teaching, San Diego, CA.

Wallace, J. & Louden, W. (1992). 'Science teaching and teachers' knowledge: Prospects for reform of elementary classrooms', *Science Education*, *76*(5), 1–15.

Wallace, J. & Louden, W. (1994). 'Collaboration and the growth of teachers' knowledge', *Quantitative Studies in Education*, *7*(4), 323–334.

Wallace, J. & Louden, W. (1997). 'Preconceptions and theoretical frameworks', *Journal of Research in Science Teaching*, *34*(4), 319–322.

Wallace, J. & Louden, W. (1998). 'Curriculum change in science: Riding the waves of reform', in B.J. Fraser & K.G. Tobin (eds.), *International handbook of science education*, Dordrecht, The Netherlands, Kluwer, 471–485.

Waller, W. (1932). *The sociology of teaching*, New York, John Wiley & Sons.

Welty, E. (1984). *One writer's beginnings*, Cambridge, MA, Harvard University Press.

White, J. (1989). 'Student teaching as a rite of passage', *Anthropology and Education Quarterly*, *20*, 177–195.

Wildy, H., Louden, W. & Wallace, J. (1998). 'School physics and the construction of social inequality', *Australian Educational Researcher*, *25*(2), 39–59.

Wildy, H. & Wallace, J. (1994). 'Relearning to teach physics: In the midst of change', *Research in Science and Technological Education*, *12*(1), 63–75.

Wildy, H. & Wallace, J. (1995). 'Understanding teaching or teaching for understanding: Alternative frameworks for science classrooms', *Journal of Research in Science Teaching*, *32*(2), 143–156.

Willis, P. (1977). *Learning to labour: How working class kids get working class jobs*, Westmead, Saxon House.

Witherall, C. & Noddings, N. (eds.). (1991). *Stories lives tell: Narrative and dialogue in education*, New York, Teachers College Press.

Wolfe, T. (1973). *The new journalism*, New York: Harper and Row.

INDEX

183

Science & Technology Education Library

Series editor: Ken Tobin, *University of Pennsylvania, Philadelphia, USA*

Publications
1. W.-M. Roth: *Authentic School Science.* Knowing and Learning in Open-Inquiry Science Laboratories. 1995   ISBN 0-7923-3088-9; Pb: 0-7923-3307-1
2. L.H. Parker, L.J. Rennie and B.J. Fraser (eds.): *Gender, Science and Mathematics.* Shortening the Shadow. 1996   ISBN 0-7923-3535-X; Pb: 0-7923-3582-1
3. W.-M. Roth: *Designing Communities.* 1997
                                           ISBN 0-7923-4703-X; Pb: 0-7923-4704-8
4. W.W. Cobern (ed.): *Socio-Cultural Perspectives on Science Education.* An International Dialogue. 1998   ISBN 0-7923-4987-3; Pb: 0-7923-4988-1
5. W.F. McComas (ed.): *The Nature of Science in Science Education.* Rationales and Strategies. 1998   ISBN 0-7923-5080-4
6. J. Gess-Newsome and N.C. Lederman (eds.): *Examining Pedagogical Content Knowledge.* The Construct and its Implications for Science Education. 1999
                                           ISBN 0-7923-5903-8

KLUWER ACADEMIC PUBLISHERS – DORDRECHT / BOSTON / LONDON